SOFT SKILLS TO ADVANCE YOUR DEVELOPER CAREER

ACTIONABLE STEPS TO HELP MAXIMIZE YOUR POTENTIAL

Zsolt Nagy

Apress®

Soft Skills to Advance Your Developer Career: Actionable Steps to Help Maximize Your Potential

Zsolt Nagy
Berlin, Berlin, Germany

ISBN-13 (pbk): 978-1-4842-5091-4   ISBN-13 (electronic): 978-1-4842-5092-1
https://doi.org/10.1007/978-1-4842-5092-1

Copyright © 2019 by Zsolt Nagy

Managing Director, Apress Media LLC: Welmoed Spahr
Acquisitions Editor: Shiva Ramachandran
Development Editor: Rita Fernando
Coordinating Editor: Rita Fernando

Cover designed by eStudioCalamar

Distributed to the book trade worldwide by Springer Science+Business Media New York, 233 Spring Street, 6th Floor, New York, NY 10013. Phone 1-800-SPRINGER, fax (201) 348-4505, e-mail orders-ny@springer-sbm.com, or visit www.springeronline.com. Apress Media, LLC is a California LLC and the sole member (owner) is Springer Science + Business Media Finance Inc (SSBM Finance Inc). SSBM Finance Inc is a **Delaware** corporation.

For information on translations, please e-mail rights@apress.com, or visit http://www.apress.com/rights-permissions.

Apress titles may be purchased in bulk for academic, corporate, or promotional use. eBook versions and licenses are also available for most titles. For more information, reference our Print and eBook Bulk Sales web page at http://www.apress.com/bulk-sales.

Any source code or other supplementary material referenced by the author in this book is available to readers on GitHub via the book's product page, located at www.apress.com/9781484250914. For more detailed information, please visit http://www.apress.com/source-code.

Printed on acid-free paper

Contents

About the Author

Zsolt Nagy is a software engineer, manager, tech lead, and mentor specializing in the development of maintainable web applications with cutting-edge technologies since 2010.

As a software engineer, Zsolt continuously challenges himself to stick to the highest possible standards. He put extra effort into building a T-shaped profile in leadership and software engineering. You can read more about his specializations by visiting his blogs. The tech blog, www.zsoltnagy.eu, is on improving your JavaScript skills by solving tech-interviewing questions and developing real-world web applications that you can monetize or display in your portfolio. The software development career blog, devcareermastery.com, is all about finding fulfillment in your career as a developer, with the help of emotional intelligence.

Introduction

In recent years, there has been a continuously increasing demand for software developers. The US Bureau of Labor Statistics predicts that there will be a 24% growth in the number of software development positions by 2026 with respect to the 2016 data.[1] This growth is predictable, as there is software in almost every device we use. Most organizations need customized software for their daily operations. Besides business reasons, we also use apps for leisure and for making our lives easier.

The need for software development is not going away even with the newest AI advances. Someone needs to design and integrate even AI services. Furthermore, AI as a service will make the lives of developers easier, making it possible for developers to concentrate on higher-level tasks. This means to you that your job may become more interesting than before, and you can add more value to your organization with the help of the latest advances in technology.

Unfortunately, for many developers, the future does not appear to be that bright. Many developers I have been in touch with since I started coaching in 2016 are uncertain about their future. It is very easy to get an average job in an environment where you trade your potential as a developer in exchange for a comfortable salary. This package makes many developers accept and ignore the moderate toxicity in the environment they are in. Many developers, myself included, assumed at one point in our careers that this toxicity is the norm. After all, there are problems in all companies, right? Sometimes we hear about fun stories about engineering culture in companies like Google or Spotify and see photos of nice offices that look like a playground. However, this reality is rarely the reality of our own; therefore, many of us accepted at some point in our careers that our situation is different. We tend to settle for what is available for us.

In my own life, I made a decision a few years before writing the first version of this book that I want to create a life for myself on my own terms. This life is free from daily compromises, toxic environments, and accepting what's wrong. If something is wrong at work, I not only point it out, but I do everything in my power to show my surroundings a way out.

---

[1] www.bls.gov/ooh/computer-and-information-technology/software-developers.htm

When it comes to my own personal life, I realized early enough that I have dreams that go beyond what a 3%–10% yearly raise can finance. Therefore, I took charge of my own career and identified skills that not only set me up for a higher growth in earning ability as an employee but also come handy once I become self-employed or start growing businesses.

Throughout this journey, I have read more than 100 books in the area of professionalism, taking responsibility, mindset, entrepreneurship, software engineering best practices, communication, learning, and peak performance psychology. Reading books did not take me far though. After all, any Internet marketer can present bookshelves behind a Lamborghini claiming that reading takes you very far in life. Applying the principles you learn from these books is what takes you to the next level. This is what I did. I identified actionable steps that helped me design a fulfilling career. This book will show you how you can do the same.

Your Expectations

Does it make sense for you to work on a product using an outdated stack? That stack may have been cutting-edge 15 years ago, but today, this choice just slows you down and gives you daily hurdles. There are options providing better developer experience.

As you are reading these lines, think about your own expectations. Why did you pick up this book?

Possibly you are **not satisfied with your current salary**. You may feel that you are underpaid right now. You may have reached a **career plateau** and got stuck there. It is an unpleasant emotion to feel stuck, because it feels like moving backward. In this book, I will show you not only how to make more money with software development but also how to add value in the process. While creating win-win situations, rewards will move you forward even if you think you are in a plateau where there is no way further up.

Regardless of your career options and salary, you might find yourself in an uncomfortable situation where your natural human need for safety and security is not being met. Many companies circulate **unhealthy tension at the workplace**. If you feel insecure about your job, something has to change. As it may take too much to change a toxic corporate environment, you may want to find a better job for yourself and let your current colleagues fight their meaningless battles for the illusion of power.

It may happen that everything seems just fine with your working environment, but you want **better working conditions**. You can get inspired by what is available out there by reading my article on finding the workplace of your dreams.[2]

You may also have a desire of **building something that lasts**. For many of us, our ultimate goal is to affect the lives of as many people as possible in a positive way. This requires an online presence and a proper personal brand, as well as some relevant experience. We will break down how you can plan the transition from working on something you don't find meaningful at all to living your purpose.

After the countless workshops in change management, I have concluded that change often comes with pain and uncertainty. Change has to be managed until the end state is congruent with who you are. If this vital ingredient is missing, chances are that you will unconsciously sabotage your own progress. I did the same thing back in the days.

My Story

My own career got an excellent start. I was a consultant at my university, and I was writing my master's thesis in exchange for a great salary funded by the EU research budget. I worked full time on researching a topic that got created as a synergy between the Web, mathematical logic, and software development. I became a speaker at conferences and worked with PhD students. Life was fun.

However, not long before I finished writing my master's thesis, I experienced a strong force that pushed me away from this safe, secure environment. I wanted something different. Primarily, I wanted to affect the lives of other people, because I didn't see that my research affected the lives of others to the extent I wanted to. Back in 2006, I couldn't think 15 years ahead; and back then, the software I wrote could only be used in research context to draw basic conclusions on large datasets. As I wanted to gather hands-on experience, I started interviewing for jobs.

I still remember my first interviews. Even though I had the best possible track record you can start with, things went a lot worse than I expected. I crashed and burned in many interviews. Sometimes HR asked me to rate my skills on a scale from 0 to 10. I was honest about not knowing everything, and I underestimated my skills in some areas, so I got eliminated. Once HR dumped me

---

[2] http://devcareermastery.com/find-the-workplace-of-your-dreams/

because I insisted in not disclosing my well-protected secret salary expectations. Even today, some popular sites suggest you not to disclose your expectations, and this blows my mind, because I had experience from 2006 about getting eliminated multiple times just for executing this trick.

Another hurdle in the interviewing process was that I thought of myself as a natural introvert, someone who is not very good with people. I got deeply traumatized by the way how literature got taught during my high school years by a tyrannical teacher who was not motivated at all. As my entry ticket for a government-financed university spot led through acing literature, my relationship with reading and appreciating art got damaged. I did everything I could to ace literature, but I never truly understood it or appreciated it. I just memorized and forgot facts. When someone was playful with words or became creative, I got shoved back into my head and became like a robot. This trait was not too helpful when answering the "Tell me a bit about yourself!" interview question.

As I got some tech-interviewing experience under my belt, I got better at answering questions that made no sense to me. I also developed a talent for making my interviewers do the work for me by making them talk about what they wanted in a candidate. I learned that feeding back their own thoughts to them was something that most interviewers appreciated. This gave me a good start, because my tech skills were solid enough to get hired. Therefore, I could choose between multiple options.

Something was still wrong. I was not enthusiastic about joining any of the companies that wanted to hire me. I believed that there is more to life than working among people who clock in at 9:00 a.m., clock out at 6:00 p.m., and spend their time demotivated, doing a task I found meaningless. I could choose between joining a consulting company and a software house, where unpaid overtime was the norm; or I could choose a smaller company, where half of my salary would have been paid illegally in cash, and my job would have been to research the documentation of standards. The guy who explained his standardization job to me was so low in energy that I couldn't believe my eyes. His only hope was that I would get hired and he could move on to do something more meaningful.

I turned all these offers down and made a bold move by accepting an offer that seemingly made no sense. I accepted an offer with low salary. From my peer group, I was the only one who saw the level of responsibility I took by accepting this position. Within a year, I became the tech lead of multiple projects.

Unfortunately, I made some mistakes with this company. The low salary was accompanied by a verbal promise of possible future ownership. The promise was kept vague, and it never became reality. Taking the offer without negotiation was another big mistake. Later I found out that even though salaries were the same among developers, one person earned more than us due to a deal he negotiated. He was less useful at work than us, so it was not because of his skill level or experience; it was solely because of his negotiation skills. I never asked for a raise even though I clearly exceeded expectations. As a result, I never got a raise.

I saw business people make ten times my yearly salary in a weekend. This planted a belief in my mind that there is more to life than employment. This is why I succeeded in creating other streams of income, let them be playing poker on the side, starting a tutoring business, or getting a remote job involving the management of affiliates.

These deals gave me even more experience, but when things went too well, I started sabotaging myself unconsciously. I knew how to earn five figures with side-income, but once I reached $30,000–$40,000, I felt the same apathy as with my research job at the university. My self-esteem calibrated me to a certain range of earning ability.

Fortunately for me, I got an opportunity and landed in Malta in 2009. As my costs of living increased, I redefined success in my mind as well. My gym membership at the Hilton and my apartment with sea view also helped sending my mind the signal that I need to find a way to become more successful.

After a year in sales and marketing as a country manager, I wanted to get back to software development, because I missed the challenges tech gave me. I got hired as a contractor for 3 weeks to finish a sports betting site. Once I was done, the company liked me so much that they offered me a full-time position, and I soon started leading a team.

Throughout my 3 years in the company, I got two raises: one to cover the inflation and one symbolic 500-euro raise. The company wasn't doing so well. However, some people around me were more assertive than me, and I found out in hindsight that they got multiple raises.

This was when I realized that something had to change. I gave up playing professional poker, which put a big hole in my family budget. This way I was committed to learning the ins and outs of negotiation, personal branding, and career management.

I always wanted to understand why something works, not just believe in concepts others tell me. I did my research and organized information in a way that enabled me to see the big picture, just like during the old days of preparing for my exams. As I continued exploring how to take charge of my own career, I realized that personal development, communication techniques, and negotiation advice only offered symptomatic treatment as a compensation that was absent in me. The root cause of most of my struggles came from within: I had low self-esteem and a distorted self-image.

This research not only allowed me to triple my own salary but also gave me an opportunity to develop skills that will come handy for me in the next 10 years. At the same time, I started recognizing the same flaws in the actions and behavior of other people. This enabled me not only to realize that I am not alone with these problems but I could also help some of them, as I was a bit ahead on the same path.

This is why I found it crucial to write down all my thoughts about how to design your career. I am still developing software, and this hobby of mine is not going anywhere. As a technical lead and engineering manager, I still have the chance of solving meaningful problems in my career. At the same time, I interview people on a weekly basis, and I see how salaries are set. This means to you that you are not reading the conclusions of a wannabe "coach" who takes money from the job center while pretending to give advice on topics he only knows from three to four books.

The city I live in has many of these coaches, and I know what it is like when someone trusts them and they lead you to a place that makes no sense in objective reality. I also know what it is like to go online, and before watching a YouTube video, you face an Internet marketer flashing a fake lifestyle, trying to convince you that his material is top-notch, while in reality, they just package three to four books and free information in a product.

This book is different. It contains my experience of having worked both in a startup and in a corporate environment, having worked as a freelancer, and having launched books, courses, and other products. I have also worked as a consultant, and I am running two blogs: zsoltnagy.eu and devcareermastery. com. This means to you that there are areas in which I can help you out, because I have experienced the same situations and went through the same hurdles that you are about to go through, provided that you choose the path of career advancement.

Depending on your skill set and experience, your potential is most likely way beyond your current expectations. Apply the principles of this book in your career, and observe the results in 5 years. Most people are surprised how much they are capable of through focused effort.

My other challenge to you is to make the process as enjoyable as possible. As you gain experience, you will have more and more options to choose from. Without any doubt, you should know that as long as you are on the right track and you act consistently, you deserve to implement a proper work-life balance.

If you would like to share your success stories with me, or you get stuck at any point, drop a mail at zsolt@devcareermastery.com.

Disclaimer

I have learned throughout my career as a software developer on how to make my own journey better and more lucrative. Advice in this book worked for me and has worked for many people I have been in touch with. Given that every person and every situation is different, even though I have the best intention of helping you, advice in this book may or may not work in your specific situation. For instance, if your company is 1 month away from bankruptcy, no advice in the world is likely to help you negotiate a raise.

The information presented in this book and on the web site of the author represents the view of the author as of the date of publication and is presented for informational purposes only. This information solely represents the opinion of the author, at the time of publishing the book. No part of the book may be viewed as consulting or legal advice. Neither the author nor his partners assume any responsibility for errors, omissions, or inaccuracies.

The Importance of Soft Skills

The Developer's Edge

You arrive in front of your office, smiling, thinking about the adventures that are just ahead of you. You are in charge of a sensitive project that will impact the lives of millions of people. Your managers back all your decisions up, and your team consists of amazing talent.

You are happy to start your work today, as you can grow so quickly in an environment like this, and you start thinking about the speed at which you develop your product. There is no one single obstacle in your way that's down to bad management, lack of communication, or the usual causes why many software projects fail. As your company allows remote work, you are considering purchasing a holiday home. Your job makes it possible for you to enjoy the Mediterranean climate instead of the long, dark, cold winter days. You are working at the right place. You are living your dream.

For some people, this picture is everyday reality. For others, this scene sounds so distant that they think they would never reach this stage in their careers. The ultimate goal of this book is to make your dream come true. This book gives you the advantage to accelerate your progress in your career.

© Zsolt Nagy 2019
Z. Nagy, *Soft Skills to Advance Your Developer Career*,
https://doi.org/10.1007/978-1-4842-5092-1_1

The path toward getting your dream job is worth pursuing, as you can not only reap big rewards at the end, but you will also be rewarded during your journey. Regardless of whether you are looking for more money or better job security or building something you are proud of, I challenge you to enjoy each step of your journey. When making choices, always remember that progress is happiness. Reaching your goals does not make you happy. Who you become along the journey is what makes the difference.

You will take your first steps by reading this chapter.

We will explore why software developer careers are special. You will find out why you might want to consider getting help from a software professional instead of a career coach who has never worked in the field of software development.

You will then learn about the four essential soft skills that shape your career. Not only this book but also your journey is based on these four soft skills. Master these skills, and you will always know what the next step is going to be from the perspective of your career.

With the help of these four essential soft skills, we will construct a holistic model giving you a map that guides you on your journey. We will then go over ways to score some quick wins in your career you can implement right away.

Never "leave the scene" without taking some action; implementing some quick changes in your career will build momentum and accelerate your progress, meaning don't just read this book and then go do something completely unrelated, such as watch a movie or play video games. Don't say "Someday I will do it" because that day will never come. Take action now. It will take you a lot less effort to work on your career advancement on a daily basis.

Let's start our journey by exploring what makes software developer careers special.

Why Software Developer Careers Are Special

Goals of this section:

- Understand why it is great to be a software developer in the 21st century
- Understand the challenges of software developer careers

If you want to make a difference in your career as a software developer, chances are you are already convinced that you have one of the best professions on the planet. As one of my friends put it, we are paid to do what we love. Crafting software is a rewarding and creative experience. We take pride in delivering professional solutions that work.

Unfortunately, software development also has a dark side. Software developers often focus on abstract concepts, interacting with computers instead of other humans. This gives us a disadvantage when it comes to expressing our thoughts and emotions clearly, as we have less experience at them compared to client-facing roles, such as a salesperson. Therefore, many of us need help to negotiate better working conditions for ourselves, or to market ourselves properly.

Career advancement can often be blocked by our lack of ability to cooperate with people who are not tech-savvy at all. To cooperate with business stakeholders, we need skills no one taught us during our studies.

After reaching a plateau in their careers, many developers start searching for opportunities to make progress. Some people improve their professional expertise, while others seek career advice from a coach who either does not understand software development in general or has lost touch with the industry many years ago.

Some coaches are entrepreneurs or developer celebrities also known as rock star developers. They often inspire us and talk about great topics. Yet, the utility of their advice depends on their expertise. I learned a lot from them about professionalism, when to say yes, when to say no, how to use some productivity tools, or how to manage teams. I have experimented with a lot of concepts, strategies, tips, and best practices; and I have found that some work for me really well, while others don't.

After learning some lessons the hard way, I asked myself the question: What if the strategies of other people are just not meant to work for me? What if I had to take responsibility to discover what works in my situation knowing my own strengths and weaknesses? What if these concepts just made me a clone of insecure people wanting to act as if they were confident?

The problem with acting confident comes when you do not feel confident. Acting confident is like a bugfix that hides the symptoms, not addressing the root cause. Hiding all symptoms of an underlying root cause often takes a lot more work than to fix it properly. In this book, you will learn how to address the root cause of feeling insecure. Once the root cause is gone, you will become confident in situations where you had to pretend that you were confident. Career coaches give us a path on a map, without considering the real terrain, and without considering our own abilities. Some people succeed, because following the path is better than wandering aimlessly. At the same time, others drown, because no one taught them how to swim, and the path leads through deep water.

Don't confuse a map with a step-by-step solution. I know it is tempting to look for easy solutions, but sometimes, easy solutions just do not exist. In these situations, taking responsibility for your actions is the only lasting solution. Solutions that sound easier are nothing else but intellectual fog.

▓ **Note** Intellectual fog is when you follow a concept that appears to guide you on the surface. However, instead of guiding you, intellectual fog distorts your perception of reality. A lot of concepts prevent you from formulating clear judgment on a situation and limit your thought process. Believing that the Earth is flat is intellectual fog. Similarly, believing that you should never disclose your salary expectations before the interviewer does is also intellectual fog. My method acting coach, Shredy Jabarin (`www.shredyjabarin.com`), introduced me to this concept. He refers to getting rid of intellectual fog as emptying the trash.

Intellectual fog occupies space in your mind and distorts your perception, because you see the world according to a filter. The majority of this book helps you get rid of this intellectual fog and allows you to see clearly.

Intellectual fog can appear in multiple forms. A strategy may not apply to you, because

1.   Software developer careers have special characteristics.

2.   Your individual situation is not compatible with the strategy.

While we will address the second point later, in this section, we will address the first point. The situation of software developers is so unique that I often had to work hard to interpret the maps and paths in a way that makes the most sense for my goals. The outcome of this process is summarized in this book.

Let's explore why you are in a unique situation as a software developer when it comes to selecting career advice that applies to you.

Demand and supply: Software developers are in very high demand. There are a lot more jobs available in the IT industry than the number of qualified professionals ready to apply. Companies don't only need software developers: they need people with a professional attitude, who are capable of cooperating with the business and can solve complex, abstract problems. Once you acquire the skills companies are looking for, you will have no problems with finding a job.

Salaries are spread in a wide range: Ranging from internships to chief technology officers (CTOs), salaries vary. As soon as you become a junior software developer, chances are you will earn more money than most unskilled laborers. As you advance in your career, you will get a chance to tap into six-figure salaries in dollars or euros.

Software developers are often viewed as a cost: This is one reason why the job market is so inefficient. IT jobs are so abstract that it requires intelligence to detect and define exceptional performance. If a salesperson closes 10 million dollars each month instead of the team average of 1 million, you know that this person is a superstar closer. What does a 10x developer do? Solve complex problems ten times as fast? Who notices that? Take architectural decisions that prevent their team from burning ten times as much money as the cost of the project? How do you measure decision quality? To this day, KPIs (Key Performance Indicators) of IT skills are hard to define.

Everyone wants to work with the best people; yet, they don't understand that the best developers want to be heroes, just like anyone else. Viewing software developers as a cost makes it very hard for developers to justify why their contribution is worth a lot more, as their performance is hardly ever tied to the profit of the company. Don't worry. The battle is not lost. I will show you how to construct your pitch.

Tension between business and IT: Unless you are in a tech company with good understanding and a healthy development process, tension may develop between your team and your employer. Businesspeople want measurable results quickly. They don't understand complexity, and they view software development as a simple linear process. If a project takes 200 days to complete, then four times the team can complete it in 50 days, right? Wrong. Fast-forward 100 days, 50% of the project is tangible and completed. Right? Wrong. Changing a sentence in the requirements document is easy; after all, it is a small change. Right? Wrong. When it comes to tension, businesspeople are often more assertive than software developers. If you give them a weak response, they will take your response as a commitment and put the blame on you for not delivering. At the same time, software developers want to be the heroes of the day, and this makes developers commit even when the associated risk is too high.

Lack of incentive to overperform: Unless you work in an innovative company that is or once was a tech startup, chances are your company does not provide you with incentives to overperform. Many companies don't even compensate you for your overtime. If you are wondering what you need to do to deserve a raise, you are not alone. I encourage you to read further, and learn how to create an irresistible pitch to your employer.

The job does not teach you to negotiate and assert yourself: Let's face it, writing code does not improve our people skills. Software developers with good people and networking skills have an easier time navigating through their career options. At the same time, salespeople close deals on a daily basis. Even though salespeople tend to earn a commission, there are a lot of things that they can negotiate. As this is their job, they are more likely to assert themselves than a software developer.

Abundance of transition paths: We have a lot of options. Some developers feel that they are forced into becoming leads to avoid a career plateau. Nothing is further from the truth. In today's world, a lot of experts earn more than their leaders. Yet, most people wonder when they hear titles like solution architect, CTO, vice president, team lead, engineering manager, design thinking coach, scrum master, product manager, technical project manager, technical lead, head of development, and software engineer. The abundance of opportunities seems to be overwhelming. If you don't know where you are heading toward, you may invest your time and effort moving toward a goal you don't want.

Your performance is hard to monitor: As your work is often too abstract to be quantifiable, people in charge of keeping track of your performance may have a hard time getting a clear view on how good you are. In some companies, people in charge of determining your salary have no clue about software development. They just see results. In order to bridge the gap between your real performance and your perceived performance, you have to make an effort to make your work visible and understandable. This is a balancing act between the dangerous extreme cases of being too shy and bragging about your accomplishments.

Having collected all these challenges, I started wondering if my original assumption is true. I assumed that software developers are happy because they love what they do. Yet, our cooperation with our employers often makes it hard to really enjoy what we are passionate about. I still believe that every developer has a chance to experience the joy of being in an environment where we can focus on our passion and not on things that bother us. I also believe that we can learn all the skills we need to get the working conditions and rewards that we want, provided that we offer enough value in exchange.

This book describes the path of learning how to create win-win situations with your current and future employers.

Four Essential Soft Skills for Career Advancement

Goals of this section:

- Find out why soft skills are important from the perspective of your career.

- Discover the four essential soft skills that act as four cornerstones of your professional career.

- Realize how these four essential soft skills help you master other skills.

Soft skills are personal attributes that enable someone to interact effectively and harmoniously with other people. In his book, *What Got You Here Won't Get You There*,[1] Marshall Goldsmith pointed out that beyond a certain skill level, everyone is good enough. Once you rise to this level, your soft skills will determine your success.

---

[1] Marshall Goldsmith, *What Got You Here Won't Get You There: How Successful People Become Even More Successful* (Profile Books, 2008).

The good news is that you can consciously focus on your soft skills. This will give you a significant advantage throughout your career. Developing soft skills will not only accelerate your career, but it will also make you a better professional.

Have you discovered any of your colleagues do something inappropriate? Have you ever had a colleague who

- Makes quick judgment on others and formalizes their judgment with destructive comments

- Tends to focus on helping anyone even when help is not welcome

- Is not willing to accept any help under any circumstances

- Keeps bragging about his or her accomplishments whenever an opportunity arises

- Spreads negative emotions at the workplace, such as anger or lack of motivation, affecting the mood of other people

- Fails to give you any feedback at any time and even fails to say thank you

- Keeps waiting for their own turn in the conversation, or keeps interrupting you, refusing to listen to you

- Is friendly with you, but trash-talks you when you are not present

- Gets offended for no reason and blames you for how they feel about a random topic

- Gets and keeps all the credit for your accomplishments?

There are countless stories about people doing things that hurt their careers. Don't be one of them!

Using soft skills, you can represent the interest of your team. Soft skills will also help you get promotions, raises, better working conditions, and increase the number of employment options by elevating your interviewing skills.

You may think at this point that it is very hard to get started. This is why we will now focus on four soft skills that will be essential from the perspective of your career. You will get a glimpse at these four soft skills to see the big picture of what this book will give you. In later chapters, we will explore each of these soft skills in detail from the perspective of your career. You will not only get some new ideas to think about, but you will also get some exercises that you can put into action.

The four soft skills are as follows:

- Personal integrity
- Taking responsibility
- Professionalism
- Communication

This list may surprise you. Where is flexibility? Where is commitment? Where is time management? Teamwork? Leadership? Mentoring? Problem solving? Courtesy?

Think about all these skills for a moment. Is it possible to exercise them without personal integrity, taking responsibility, professionalism, and communication skills? Is it possible to lead without communication skills, integrity, taking responsibility, and showing a good example by being a professional? Is it possible to solve meaningful problems in practice or mentor others if you are not a professional and you can't communicate your own thought process?

Good character traits and valuable skills are built on solid foundations. Once the foundations are there, the rest easily falls into place. Think about fixing a bug in the code. There are symptomatic treatments, and there is an approach to fix the root cause. A professional addresses the root cause. A professional keeps personal integrity by communicating that the root cause has to be fixed, even if the symptom has to be addressed before in a hotfix.

Employers look for people with these four skills. A software professional with personal integrity, self-responsibility, and good communication skills is hard to find. This is going to be your advantage. If you exercise these skills on a regular basis, your current employer may recognize changes in your attitude relatively soon. Alternatively, if your employer does not notice anything, it is not a big problem, as you will have a lot of other options to choose from. Other employers will definitely show their interest in your services.

Let's examine the four fundamental soft skills in detail.

Personal Integrity

Whatever you do in your career, always think about the big picture. All decisions have a knock-on effect on your career. Approaching your decisions with total integrity means that you will never regret anything you do.

Approaching decisions with integrity implies that you are honest and stick to your moral and ethical values. You do what you think is right. You have a moral code; and you are not willing to enter into compromises because of fear, seeking to meet expectations of others, or the desire to get short-term benefits.

Having personal integrity does not mean that you don't have to develop a persona to represent the interest of your company, for instance, in the boardroom or during negotiations. It just means that this persona is integrated with your own personality. Therefore, your actions will seem authentic and more trustable.

Integrity is the core of our holistic model. This is why you will start with developing your personal integrity in Chapter 3. You will learn how to build your self-esteem, how to develop your confidence, and how to grow as a professional. This will allow you to step up and do what you think is right, instead of seeking what other people think is right.

If you have doubts about this process, I can understand you. My actions used to be driven by my own fears. I had a lot of limiting beliefs in place that prevented me from acting with integrity. In my own thoughts, I did not deserve to do what I really wanted to do because taking care of others' needs and expectations comes first. Back in the days, this mindset not only poisoned my thoughts, but it also drained my energy and internal drive.

Whenever I visit a safe city, I have a habit of taking public transportation on Monday morning to observe the mood of people. I often see low energy and lack of motivation. For many people, their work does not give meaning to their lives. They work 5 days in exchange for having fun for 2 days. On Monday morning, the fun has just ended.

Living with integrity means that how you do one thing is how you do everything else. Therefore, you have no reason to feel any better or any worse on Monday than during the weekend. You must have seen at least one colleague of yours who comes in happily almost every Monday morning. Is it likely that this person acts with integrity? The key to job satisfaction is being able to act with integrity. You may not be there in your career yet. This is why we will work on setting meaningful goals for you and put you on a career path worth pursuing.

All chapters of this book build on integrity. Lies or any behavior that challenges your personal integrity should have no place in a life of a software professional.

Taking Responsibility

Some companies get away with continuously shipping software without a Quality Assurance team. Some companies get away with working effectively without employing project managers taking charge of coordinating resources. The necessary condition for this to work is taking responsibility on individual and team level.

Employers seek people who go the extra mile. People taking charge will never tell you that it's not their job to help their own company with an effort they are not specialized in. People who take responsibility will never sit in front of their computers feeling blocked and waiting for an answer to arrive. As the name of this soft skill says, they take responsibility by standing up and getting the piece of information they need. They never settle until they unblock themselves.

If you have the reputation of taking responsibility, good employers will value your contribution. In the era of increasing demand for software developers, many developers think that they can get away with not taking responsibility, because they find a job anyway. While some developers may get away with this mindset, chances are they block their own progress at the workplace. At the same time, you can take responsibility from day 1, and chances are you will be the first to consider for a promotion.

Some people tend to blame others for their own mistakes. In most cases, those people have a hard time admitting that they made a mistake. Your weapon will be to act with integrity and take responsibility even for your mistakes. After a while, you will realize that making mistakes is essential in growing. If you don't make mistakes, you learn slower. If you are afraid of making mistakes, your contribution will be less valuable.

Notice that personal integrity will help you when others blame you. As long as you act with integrity, blaming you won't succeed.

In Chapter 3, you will learn to start taking responsibility by introducing the idea that you are responsible for developing your skills. You will learn how to shift your mindset to grow rapidly. Later, you will discover the connection between motivation and willpower. You may draw important conclusions about motivating yourself.

Chapter 5 helps you use the power of feedback to work on your weaknesses. If you get no feedback, no problem; you will learn how to take the initiative and build feedback loops. If you aren't noticed, you can take responsibility and accomplish things that no one will overlook. Even if feedback does not result in an immediate raise, remember money catches up with taking responsibility in the long run.

In the negotiation phase of the hiring process, you can show that you take responsibility without explicitly bragging about it. You can even take responsibility for researching your salary and respond with your expectations without submitting to the intellectual fog about information asymmetry and the "first person naming a number loses" perceived axiom. As with every half-truth, the first person naming a number sometimes loses, other times wins. Everything depends on the context.

Note   Information asymmetry in the hiring process describes the lack of salary-related information in possession of the candidate with respect to the employer.

Professionalism

Professionalism is essential for career advancement. Coupled with personal integrity and responsibility, other people will view you as someone whose professional opinion is to be respected. Very few software developers master these skills. This is the number one reason why developers are afraid of interviewing and salary negotiations. This is also the number one reason why many developers lack the self-esteem required to even think about targeting one of the most lucrative positions available to them.

What is professionalism? This is a very hard question, as many people have many different criteria toward professionalism. We will explore it from multiple angles.

A professional is someone who sticks to defined high standards when delivering a service. These standards include the code of conduct of the professional.

Many people, especially outside the field of software development, just scratch the surface, and determine the level of your professionalism based on your *appearance*. Although appearance is important, it is not everything. Your dress code, the way how you organize your tasks, and the way how you eat, behave, and talk have an effect on how other people judge you. These aspects are all included in your *personal brand*. However, if your colleagues wear T-shirts and shorts, no one asks you to change your style and start wearing suits. Use common sense, and dress just a bit better than your colleagues. This way, they will not only accept you, but they also look up to you.

On some level, clothing has really nothing to do with professionalism. However, as a trader friend of mine has put it, "the market is always right." Rebelling against how other people perceive you doesn't make sense. Even if assumptions are false, if you dress worse than your colleagues and there are 20 empty bottles of soft drinks on your desk, many people will view you as less professional than others.

The second component of professionalism is your *qualifications, degrees, accomplishments, and certifications*. Let's face it, a university degree will mean more to some employers than a degree obtained in a no-name college or bootcamp. Standard certifications may also be important for large companies with established processes.

The good news is that in the 21st century, you can make a difference even without a degree or any certifications in your hands. If you have a professional attitude, you can solve meaningful problems. If you know how to establish your personal brand and market it, you will reach your career goals faster than by collecting certificates. Well-marketed hands-on experience almost always wins over theoretical knowledge. Furthermore, many prestigious companies don't even look at your qualifications or certifications; they just evaluate your portfolio and experience.

We will work on all of these components in Chapter 4. You will read what it takes to create a professional application package, and you will also get advice on establishing your online presence with little work.

Little work does not mean that you plant a career seed and your online presence will magically grow out of nothing. Little work means that we apply the quadratic Pareto principle. According to the Pareto principle, 20% of your efforts bring 80% of your results. Applying the same Pareto principle inside the 20%, we can conclude that 4% of your efforts bring 64% of your results. We will identify the quick wins that will make a big difference in your career. For instance, in Chapter 2, you will dream big, and create a clear career path for yourself. This includes finding a specialization worth pursuing as a software professional.

In Chapters 6 and 7, you will learn the ins and outs of researching your salary to avoid looking unprofessional when asked about your salary expectations.

Being competent in a specialized area is important, and it characterizes software professionals. Other people will come to you for advice, and they will trust your abilities. In Chapter 4, you will read about creating a learning plan for yourself, and apply project-based learning and lateral thinking to accelerate your learning process. Effective learning pays off. Many people have a hard time keeping up with technological advancement. They keep reading books and never stop for putting their knowledge into practice. When it comes to learning, professionals know that less is more.

One important aspect of professionalism is that you are reliable and accountable. Instead of overpromising and underdelivering, you do your best to keep your promises; and if these promises cannot be kept, you inform everyone as soon as possible. Instead of hiding behind an excuse, you hold yourself accountable, and you are available to discuss conclusions of the mistake. Professionals reserve the right not to be perfect, and they do their best when it comes to correcting their mistakes.

Accountability and taking responsibility go hand in hand. A model including responsibility in professionalism would also be valid. However, in this book, we will separate responsibility from professionalism, because you can take responsibility in other areas of your life, not only your professional career. Professionalism does not require you to process traumatic events that block you from performing well at the negotiation table. Taking responsibility does, because it implies that you refuse to be a victim of circumstances and you refuse to accept the status quo.

Integrity means that you live according to your values. The values important to you are included in the professional's code of conduct. As long as you act according to your professional code of conduct, you act with professional integrity. Once your professional persona is integrated into your own personality, you start acting with total integrity.

As an added bonus, there's a whole section on professionalism in Chapter 5. Building on the foundations of what other authors have written about the topic, we will create our own professional code.

Communication

Our fourth and last essential soft skill is vital regardless of what we want in life. Communication not only complements all other soft skills, but it also enables you to take them to the next level.

Many software developers have a hard time expressing themselves. Sitting in front of a computer several hours a day and writing code, thinking about abstract ideas, and exercising your analytical mind does not help you much in communicating better.

Communicating with others is not the only task that you should learn to be more successful. We will also focus on communicating with the person you spend the most time with in your life: yourself. Things you tell yourself determine your self-esteem, shape your self-image, and can make you more confident. You will not be driven by your fears. You will allow yourself to grow beyond your current imagination. This is what Chapter 3 is all about.

In Chapter 4, we will establish your online presence. This is a written communication task, where you have to convince the external world that they should consider you. Presenting yourself is a vital skill if you want to target better jobs.

Communicating at the workplace determines how much your team will trust you. Giving and receiving feedback determines how quickly you can grow. Feedback also increases your chances at the negotiation table, when it comes to talking about your next raise. Chapter 5 gives you the strategies you need to establish the fundamentals of effective communication.

Based on solid foundations, you are ready to negotiate a higher salary for yourself, or start your job hunt. In Chapters 6 and 7, we will fine-tune your negotiation skills, and make your life as easy as possible. You can become a confident negotiator. This includes your ability to face your current employer as well as other employers and assertively represent your interest while creating win-win situations.

I have a zero-tolerance policy toward intellectual fog. The market tends to give you easy solutions to problems where easy solutions are not meant to exist in the first place. Unfortunately, it is harder to market the idea to the masses that taking responsibility and improving your communication skills lead you to where you want to go to, and this is the easiest solution that covers your needs. Any shortcut may just create an illusion of progress and distort your perception of reality. Oftentimes, the worst thing that can happen to you is if you succeed with a shortcut.

When it comes to communication, most people give you universal rules such as "the first person naming a number loses at the negotiation table." This theory is backed by information asymmetry. I have evidence to back up that in the path I am laying out, not naming a number first may more often backfire on you than you would think. It does not mean that it is always beneficial to name a number first, because some corporate environments may require you to play the negotiation game taking information asymmetry into consideration. However, assuming you advance to the top 5% of your profession, oftentimes you have enough leverage to bend the rules and earn more money.

This book is all about giving you the foundations that will enable you to detect intellectual fog and discard it.

Most people put themselves in very tough spots by relying on negotiation techniques, sacrificing their own personal integrity. We will work on making your life as easy as possible. You do your job well, and your employer will often back you up. Otherwise, your current employer will be likely to lose you, as other potential employers will fight for you.

I also have a zero-tolerance policy toward manipulation, lack of honesty, narcissism, Machiavellianism, psychopathy, and blackmailing. Although having a BATNA (Best Alternative to a Negotiated Agreement) is always helpful, you don't need an external offer to negotiate a higher salary with your current employer. Once again, in some environments, people will throw money at you to stay, but a professional's skill level is often sufficient as a BATNA. In other words, if you follow the path I am laying out, your employers will know that you can walk away and find a great job on your terms at any time, and it is their interest to keep you satisfied.

You won't need to brag about your accomplishments, as you will learn how to use your feedback to your advantage and how to earn the respect of your managers so that you can see and hear your accomplishments from them.

Although this book focuses on mostly healthy environments, there are times when you have to defend yourself from toxicity. Notice that these four essential soft skills already put you in a great position, giving you some armor against predators who are fueled by consumption and are interested in gaming the system for a self-centered purpose. For instance, notice that communication and personal integrity include defending your own psychological boundaries.

Therefore, others won't be able to violate your boundaries for making you responsible for how they feel, especially if their own emotions are disconnected from objective reality.

Consequences of the Four Essential Soft Skills

Integrity, responsibility, professionalism, and communication help you improve other areas of your professional career. Let's see some examples.

Teamwork: Professionals realize that they have to rely on other people. By taking charge and communicating effectively, you will be able to cooperate with your team better, for the purpose of solving meaningful problems better and faster than alone.

Work ethic: The code of a professional includes high standards when it comes to work ethic. Otherwise, the professional risks losing their integrity. Professionals learn how to work smart, take the initiative, and are able to cooperate with others when they need help.

Leadership: By showing your professional attitude and taking responsibility, you encourage your team to adopt the same traits and become better people.

Mentorship: Software developers look for mentors inside and outside the workplace. You can be a credible mentor even as a junior developer. At least you have a perspective of telling others how you have overcome the difficulties of learning something new. A senior developer can hardly recall past struggle with learning. You can teach others how to come up with professional solutions. Mentoring others is also a great way to fine-tune your own communication skills.

Problem solving: The attitude of a software professional continuously challenges your problem-solving skills, as you keep working on meaningful problems in your career. A professional never settles for a symptomatic solution. A professional always digs more deeply to find and eliminate the root cause of a problem. Assuming that you get better at whatever you do on a regular basis, developing your problem-solving skills will be a side effect of professionalism and your integrity of wanting to become a true software professional.

Time management: Another skill that seems to have little to no correlation with the four essential soft skills. Let's dig a bit deeper. Topics covering time management, productivity, deep work, and effective work are vital in the life of a software professional. Assuming that you want to deliver a professional solution, it makes little sense to waste your time on things that don't matter from the perspective of your career. It is also your responsibility to set your boundaries. Sometimes you have to tell other people that you are in the middle of something and ask them to come back to you in an hour or two.

This requires communication skills. Sometimes you have to tell yourself that you need to stand up, walk, relax, and regain your energy. Other times you may have to tell yourself that browsing Facebook won't solve the problem that you feel bored. Maybe you need to solve the cause why you are bored, and make your work more effective.

Emotional intelligence: This is the last, but not least, important example of how the four essential soft skills can be applied to improve your career. The more you develop your emotional intelligence, the better you become at the four essential soft skills. Therefore, you can use most of this book as training on emotional intelligence, in the context of software development.

In this book, we will use these four soft skills to create the developer's edge. This evident advantage will not only make you a better professional, but it will also help you get your dream job. Let's continue with the holistic model of our approach.

Your Path Toward Your Dream Job and Dream Life

Goals of this section:

- Observe your journey from a holistic perspective.
- Learn about the importance of each step to see the role of the upcoming topics one by one.

The task of the holistic model is to show you the part-whole relationship between steps in your journey. I will show you how different chapters of the book will give you the results you want.

Similar to the layered ISO-OSI model in computer networks, we will think in layers. We will work on three simple layers that will give you the most benefits. These three layers are as follows:

- Career mindset
- Expressing your identity
- Negotiation skills

Career mindset, the first layer, is all about emotional intelligence. You will learn how to build your self-esteem and form a proper self-image to become unstoppable. You will be able to accelerate your growth by managing your emotions, and allowing yourself to make mistakes. Instead of running away from your fears, let them be public speaking or a tough conversation with your boss, you will learn how to face them. You will also develop methods to motivate yourself even during tough times.

We will develop all four essential soft skills by building your career mindset. Personal integrity raises your self-esteem. You will think about yourself as someone who deserves to be successful. You will learn how to eliminate negative emotions to keep your professional attitude even when others give up on staying professional. Last, but not least, your emotional balance allows you to communicate effectively, without tricks and techniques. Once we lay down the fundamentals of your career mindset, we will move on and design your career path. You will have all the tools you need to dream big. Dreaming big is essential in designing a 5-year career path for you. This will be your internal drive, your fire when it comes to getting up and working toward your goals.

The second layer gives you tools that help you **express your identity**. After finding out what you want from your life, you will build your online and offline presence. You will learn how to design your personal brand. You will be able to create better marketing materials for yourself than many freelancers or even some companies. If you want to become a rock star developer, you will be able to reach it. If you want to raise your salary to six figures, you will be able to reach it easier with a proper personal brand, resume, online portfolio, blog, workshop or conference presentations, and other materials.

You will learn the ins and outs of effective communication, to represent your own interest, while taking the interest of others into consideration. You will develop your professional code of conduct, your rules that define your boundaries as a software professional. You will also learn how to give and how to take and encourage feedback that may even help you get promoted without negotiation.

The second layer of expressing your identity strongly relies on the first layer. If there is no mental balance, you will lose your integrity while trying to express yourself. Assuming your career mindset is there, your attempts will simply make more sense, and you will create more impact.

Based on these two layers, the third layer representing your **negotiation skills** will become very simple to master. Many information products make the evident mistake of focusing on negotiation skills right away. In other words, they give you a pill to treat some symptoms, that may or may not bring you forward. By working on lower layers first, it will take you a lot less effort to focus on your negotiation skills.

Other people will know that you have personal integrity, you are a professional, and you take full responsibility for your own actions. Your personal brand and your attitude will speak for themselves.

I can understand that the two chapters on negotiation skills are the most popular topics in any career book. This is how humans work by nature. I can hear the argument, "Yes, Zsolt, this mindset thing is cool and all, but I first want to make money, then I will focus on the other two layers." The problem is that knowing and researching the tactics is easy. Using these tactics however requires that you take responsibility and build the foundations.

Most quick solutions are like the red pill in *The Matrix*. You just swallow it and you will see. This pill does not exist in real life, except if you call it taking responsibility for increasing your competence. Most people fail because of outsourcing responsibility to an external authority. As you are reading my book, I could be this authority for you, and I could make you believe all you need is a script and a set of actions to negotiate a higher salary for yourself. Some of my readers would succeed. Others would execute the steps with elevated heart rate, feeling shame and guilt in the process.

On the other hand, if you implemented all the changes in the first five chapters and did not even read how to get a raise or a new job, would you have good chances to drastically increase your earning ability? For many of my readers, the answer will be yes.

I can still highly recommend that you read the two chapters on getting a raise and interviewing for a new position. Regardless of how professional you are, there are some insider information that will increase your chances. You will signal your interest at the right time, you will use the right salary ranges, and you will demonstrate your skills in the right way. You will also learn what you don't need to succeed and what popular career advice decreases your chances at the negotiation table.

At the end of most sections, some exercises will challenge you to put the ideas in this book into action. These questions will not only encourage you to think about your career, but they will also help you improve your skills.

I hope you are now excited to begin your journey. Who knows, you may have purchased this book to negotiate a 20% raise. This is just the beginning of the transformation that is ahead of you.

There is no better way to start this journey toward mastery than accumulating a couple of quick wins. You will make measurable progress in the course of days if you implement some ideas from the next section.

How to Build Momentum with Quick Wins

Goals of this section:

- Free some time to spend on tasks that matter to you.

- Become more organized and show it to your lead.

- Develop a learning plan that builds your online presence.

- Find out how to research salary ranges, companies, and interview questions.

- Use your research to write a cutting-edge application package that won't be rejected.

If you find yourself stuck in a deep hole, you need conscious effort and a pain period associated with this effort to get out. This section will make this pain period as easy and straightforward as possible.

Note   Toxic self-help paints the picture of a perfect role model who has full emotional control and all opportunities. These role models often attribute their own success to making a decision to perform at their peak, hustle, grind, become mentally tough, and achieve. Followers of these role models often buy into this dream and start imitating their actions without putting proper systems in place. This results in temporary highs, in exchange for a continuous drain on willpower. As their willpower is depleted, previously enthusiastic followers give up on becoming the person they dreamed to become and settle for a comfortable life.

Gurus, role models, influencers, and celebrities preaching actions that lead down the achiever's path forget just one thing: empathy toward the everyday reality of a regular person. You can only conquer the castle at the top of a hill if you first climb out from the deep hole you find yourself in. Trying to make major differences in your life is like shooting dirt at the castle from your hole using a slingshot. Eventually, you get exhausted before accomplishing anything.

The first steps are often the hardest to take. One of my peers from university, Brian, wanted to launch a personal project, and asked me how I had energy to run two blogs next to a full-time job. I asked him to let me know about his day.

He usually wakes up at around 5:30 a.m. and commutes 45 minutes to work. If he ever sleeps late, his commute becomes 1.5 hours because of the rush hour. He normally finishes work at 5:00 p.m., except if there is an emergency situation. In the past 2 weeks, he typically got home by 8:00 p.m., because he frequently had some errands to do after work. Normally, however, he gets home between 6:00 p.m. and 8:00 p.m.

"When do you go to sleep?" I asked him, while noticing he looked tired and exhausted.

"Around midnight. But I barely have the energy to do anything productive between 8:00 p.m. and midnight," he added.

"You need at least 7 hours of sleep unless you optimized your sleep with some scientific methods. Some people need 8 hours. No wonder why you cannot focus in the evening."

"I know, but my wife usually goes to work at 9:00 a.m., and she cannot sleep before midnight. I have a hard time getting to sleep too."

"Do you meditate or exercise regularly?"

"I run for half an hour twice a week. But besides that, no."

"What about the weekends?"

"Sometimes I sleep until 11 and I feel tired. I can still do some work on Saturday, but on Sunday, we have some programs with our family."

The diagnosis is clear. Brian made life choices that do not suit him, and exploits most of his energy reserves. By the time he gets home, he cannot even focus.

Shape Your Environment

After a few months, Brian made a choice to move into the city. As he could just terminate his rental agreement in a few months, he had an easy time moving. This change allows him to start work at 9:00 a.m., sleep until 7:00 or 8:00 a.m., and stop compromising his sleep schedule.

Until the move happens, Brian asked for work from home a couple of days a month, and he got the approval. Saving a long commute reaps rewards.

In your case, there may be other environmental constraints that do not enable you to perform at your peak. Find them and shape your environment accordingly.

He also found a gym and booked nonnegotiable time for his exercise during the day, just before lunch. To regain even more energy, Brian found some alone time for a daily 15-minute meditation. While others went on cigarette breaks to pretend that they relax, he decluttered his own mind and truly relaxed with meditation.

Making some of these major changes may not be possible for you; therefore, you can find other ways to save time and gain more energy: in the next action item, you will find a resource to save 1 hour a day.

Save Time by Gaining an Hour a Day

As software developers, we tend to struggle with time. There's never enough time in the day to do everything you want! What if there was a way to make time? Would you spend at least a part of that time on your career more happily? Of course you would. There are some techniques you could try to make yourself more efficient so you could spend time on other things. For tips on reclaiming time in your life, go to my web site (www.devcareermastery. com/onehouraday) to download my guide and implement this quick win in your life.

Increase Your Conscientiousness by Becoming More Organized

Conscientiousness is one of the big five personality traits that positively correlate with success at the workplace.

■ **Note** The big five personality traits, also known as the OCEAN model, consist of Openness, Conscientiousness, Extroversion, Agreeableness, and Neuroticism. The article https://positivepsychologyprogram.com/big-five-personality-theory/ explains this model in detail.

Most software developers have to fulfill errands. Write an application, fix a bug, deploy an update until Wednesday, go to a meeting, introduce architectural changes, mentor your colleague, or present your solution to the board. Some tasks are urgent, others are important, and others are urgent and important. Use a software to keep track of your tasks.

Some people use simple text files to keep track of their to-do lists. As a developer, it is easy to find the flaw in taking local notes: if your computer shuts down and you don't save often, your data are lost. Furthermore, your user experience will not be optimal either, as the list is one dimensional. We need a persistent multidimensional storage for data.

One free software I can recommend is Trello.[2] Trello boards and cards let you organize your week. Imagine there are five columns for the next 5 days of the week, and a Done column, containing the tasks you are done for this week. It just takes a minute to set up a Trello card. You can add labels to indicate urgency and importance. Order your daily tasks in terms of priority. Pick a task, work on it. Once done, put it in the Done column. If someone interrupts you with a request, you can simply create a new card and insert it in the right order, based on urgency and importance. You can then say, "I will get back to you on Thursday. Is that all right with you?" If Thursday approaches and other tasks have changed your priorities, Trello reminds you that you should get back to your colleague to reschedule the task he asked for.

As a day is passed, you can move its column to the right of the board. You will reuse your column for next week. As a small change, you can also add a Today column next to Done. If today is Wednesday, for instance, put all cards from the Wednesday column to the Today column, and drag the Wednesday column to the right. If there are too many things stacking up for today, reschedule and distribute the cards based on your workload.

---

[2] https://trello.com/

Feel free to customize the board based on what makes sense to you in your role. For instance, if you need to plan ahead more than a week, add a "Later" column on the right of the board.

Your level of organization will not only save time for you. It will also help you preserve your willpower, as your decisions on what to work on will be a lot easier. You will avoid frustrations of not following things up, and you will also look more professional than without it.

Getting things done helps you build momentum. It is emotionally rewarding to finish your most important duties fast and focus on the rest of your day without worrying about anything. Sometimes you complete all your obligations by 2:00 p.m. This means to you that you are on a "freeroll" for the rest of the day. Other times you get stuck with a task. But hey, you get stuck with the most important task, the one you should be focusing on. So you have no reason to feel guilty about it. All you need to do is review the rest of your tasks and reschedule your to-do list.

Besides Trello, another cool software is KanbanFlow.[3] As long as you use it for yourself, it is free. There is a built-in Pomodoro Tracker coming with the app, so you can use it together with the Pomodoro technique that is mentioned in the guide on getting back an hour a day (www.devcareermastery.com/onehouraday).

If you need more guidance on tracking your progress and forming habits, the Habitica[4] app gives you a gamified experience on reminding you on what you want your life to be about. Habitica will remind you of elements of your daily routine by properly setting up rewards you can earn in a role-playing game virtual world. I tested Habitica for a few weeks, and the administration overhead was too high for me. However, if you are obsessed with leveling up in a role-playing game, you can pretend you are playing *Fallout 4* while working on your professional career.

There is a chance that Habitica, KanbanFlow, or Trello won't work for you, because you prefer a lower overhead solution. In this case, I would rather encourage you to look for more focus instead of micromanagement. Clarity is power. To me, the number one deciding factor on focusing better was when I limited the number of goals I pursued for any given week to three. I could only add more goals once I was done with all three goals I had set for myself. This way, I was neither frustrated by setting the bar too high, nor was I distracted. I knew what to focus on on any day, and I was intrinsically motivated to contribute on the most important task at hand.

---

[3] https://kanbanflow.com
[4] https://habitica.com/

All I did was post three goals in a mastermind group in plain text format, for example:

1. `[x] Finish Lesson 3 of my JavaScript video course.`

2. `[ ] Create the slide deck for the Puppet, Terraform, and Bamboo webinars.`

3. `[ ] Complete my tax return and prepare every deliverable for my accountant.`

Notice the simplicity of these goals. They are binary. Even though I have to deliver three webinars in the second goal, I can only tick the box if I am done with all three. Keeping things simple makes you focus. Looking at your goals primes your mind into asking yourself, "What can I do right now to make progress?" Limiting your goals to three per week primes your mind to ask yourself, "What is the most important thing for me that makes the biggest difference?"

Note Priming is the act of using the reticular activation system (RAS) of your brain to notice and interpret events in favor of taking the right action. As an example, check out https://tinyurl.com/career-priming.

The tax return has to be done, no doubt about that. Even if my accountant helps me, I have to hand in the deliverables myself. Unless I make it a priority in my life, chances are it will be 80% done on week 1, then 90% on week 2, and 95% on week 3; and I would feel burdened by the continuous context switches. Making it a priority for myself eliminates these context switches.

The same holds for employment. You may have quarterly goals. Design your week and plan what you are going to do. When someone else has a plan for you, look at your list, and think about whether that plan is more important than the goals you set for yourself. You may need your manager's help in this, so you can simply involve your manager in your plans. This leads us to the next point...

Send Reports to Your Manager

Now that you have your cards on the Done column, before archiving them, it is worth using the information for another purpose. Send your lead a simple, lean, short weekly report.

Collect your achievements from the current week, and summarize them in an email, writing a couple of bullet points about your contributions this week.

If you have open questions, include them in the report. This will ensure you will get answers, and your lead will be aware of things that are blocking you.

Near the end of the week, you will already have some cards in the columns belonging to your next week. Summarize them in a couple of bullet points under the section name "Plans for next week."

As a result, your lead will not only know what you worked on this week and what questions you are facing with, but you will also provide accurate information on your workload for next week. If your priorities are wrong with respect to the needs of the business, your lead will have a chance of reorganizing them.

Do you think that a report like this may improve your relationship with your lead? I would guess so. If your lead is not interested in such information, it may be a very good sign that you are with the wrong company.

Note that some freelancers have reported to me in the past. Some of them chose to send me a daily report with their accomplishments, blockers, and questions. I found this report the best tool to know that my budget on the freelancer had a positive return.

Research Your Salary

This book is about career progress. If you want to make progress, you have to know what your contribution is worth. Research your current position as well as your targeted position.

There are two excellent resources for making salary research: Glassdoor (www.glassdoor.com) and PayScale (www.payscale.com).

You can not only see salary ranges but also reviews of companies people work for in exchange for a given salary. Be aware that reviews are sometimes skewed. A person fully satisfied with his or her company is often less likely to post a review than someone who just decided on quitting. A satisfied person tends to see things objectively. A dissatisfied person tends to see things very subjectively. Finally, the opinions of people are limited to their knowledge. In the case of some companies, I have read many unfair reviews that were solely based on a lack of support for toxic behavior, while other companies encourage positive reviews to rank higher in attractiveness.

Another useful resource that lets you shop for jobs in your field of expertise is http://stackoverflow.com/jobs. In general, on Stack Overflow, you get job offers that should be relevant to you. Some job ads specify their salary ranges. You will not only know the salary range you should be targeting in your desired country or city, but you will also know the expectations of the company you are applying for. LinkedIn Jobs (www.linkedin.com/jobs/) may also be an interesting source of positions.

You are one Google search away from other job sites. There may be sites specializing your city or your niche.

Do not get disappointed if you are not earning the amount you should be earning. Focus on what matters, and rest assured, with the right actions, your salary will be increased.

Create a Learning Plan

Following your job research, you may have collected a couple of skills that need polishing. If you are targeting a raise or a new position at your current organization, you can also inquire from your colleagues and your leads about what would bring your team ahead. Alternatively, if you are targeting a different team, you can easily find out what skills they use on the job.

Reading about these skills is hardly enough, as you forget 90% of it right away. Taking notes adds some value, but you will still forget a lot of things. I encourage you to apply **project-based learning**.

Create a meaningful side project, requiring the skill you want to learn. Make sure your side project is not just a dummy application with no usability, but you are solving a meaningful problem.

By working on meaningful side projects, you will not only say that you have the skills required for a job, but you can also demonstrate them. Sooner or later, people will contact you with opportunities for cooperation. GitHub open source repositories may also attract other collaborators, assuming you know how to let the world know what you are developing and you find a receptive community.

Be warned though. I have seen many people build side projects that make little sense. Don't reinvent the wheel; make something unique. For instance, before React, Angular, and Vue emerged as the winners of the framework war, everyone wanted to invent their own framework. After a while, the community became skeptical about new frameworks.

If you don't have the time and dedication to create a side project, I encourage you to write blog posts. You don't even have to host your blog if you don't want to. Services like GitHub Pages and Medium can host your blog.

Writing about topics you learn is a great way for creating lasting connections in your brain. As an added bonus, you can showcase your side project or blog whenever you apply for a new position.

Review Your Resume

If you want to apply for a new job, look at the description of the position. Then look at your resume. Are the skills and experience required for the job clearly visible in your resume? Make them stand out!

Are there skills, descriptions, and qualifications that have little to no correlation with the position you are applying for? Eliminate them from your resume, or make them shorter!

Have you linked a portfolio site, side projects, blog posts, and other activities that matter for the job? If not, review them.

Are you writing too much? Not many people will read a ten-page-long resume; make things more compact. If you would like to describe your accomplishments in detail, I suggest creating an attachment, where you go into more detail. People who are interested in more details about your career will read your attachment. People who are not interested may also view your attachment as an added bonus.

Do you write about yourself (functional perspective), or about the way how the company benefits from your services (utility perspective)? I suggest doing the latter; think about what your target company wants to read.

Does your resume reflect your specialization well enough? If not, make it stand out.

If you included a photo in your resume, go to Photofeeler (www.photofeeler.com/) and test it out. If your score is bad, leaving the headshot out is better than keeping it. Alternatively, you can visit a photo studio for better headshots.

Research the Interview

The advantage of glassdoor.com is that you can research typical interview questions in your area. Many people don't know what they are doing when interviewing people. Therefore, expect weird procedures. Your chances are still high as long as you preserve your professional attitude and fine-tune your problem-solving skills.

The blog of Toptal (www.toptal.com/) also gives you some job interview questions broken down per field of interest. I personally find the interview process of Toptal fair and credible.

Don't view Toptal as the ultimate cheat sheet for cracking the code interview. Don't view books with similar titles as an artifact giving you an unfair advantage at the interview. These resources may help you, as long as you use them in the right way: to establish feedback channels to your study plan. However, no resource does the work instead of you.

If you don't know the answer to a question, instead of memorizing it, refer back to your study plan, and make connections between the knowledge gaps you have and other areas that you may already know, or areas you need to practice.

The more established the company, the more information you can find on the recruitment process and the expectations of the company. For instance, at Amazon, their internal recruiters send you a package that contains everything you need to know to prepare for the interview.

Practice the Interview

You may also apply for jobs and get some tech-interviewing experience under your belt on a regular basis to gain some practice. You can interview with a company even if you don't want to get hired by them. The good thing about interviewing often is that you are exposed to more questions and answers for free. You can use experience gained to refine your study plan. The bad thing about interviewing for the purpose of getting feedback is that you may draw the wrong conclusions about why you were not accepted.

For instance, before I started pursuing leadership and expert positions, my interviewers rarely gave me feedback. Sometimes I was not a good cultural fit. My homework assignment was not good enough. My skills were not a good fit for the company. But why? Oftentimes, it is easy to draw false conclusions and change something that is already working.

Think of the interviewing process like a funnel. If you rarely get invited to first interviews, you have to work on your personal brand to attract more opportunities. Once you get first interviews, you need to analyze everything that happens in context. For instance, you may know exactly what you are looking for. If you are clear about your values, you may be more polarizing in the interview. This means that you turn your interviewers off if they do not share your values. A rejection from this source may actually be interpreted as a success. If you got eliminated because of a gap in your competency, learn from it. If your homework assignment is not good enough, learn from it.

The hardest part of the funnel is the bottom, when you get an offer and your negotiation attempts do not succeed. You may be applying a wrong strategy for negotiation. Your performance may leave some doubts in the mind of one of your interviewers. But which one? At what stage? This is when targeted feedback helps you out more than anything else.

There are many open questions at this point, and it might be a good idea to talk through them or practice them. One way to do so is to go through a practice interview. If you're interested in having a simulated interview with me to practice your skills, feel free to contact me on my web site at `http://devcareermastery.com/free-interview/`.

Summary

We have covered a lot of foundational work in the first chapter of this book. First, you have found out why software developer careers are special in a sense that generic career-building advice may not apply for you.

In order to lay down solid foundations for your career advancement, I defined four essential soft skills that help you structure your effort in a lasting way. By sticking to personal integrity, professionalism, taking responsibility, and developing your communication skills, you will have solid foundations for adding value at your workplace, and eventually, your earning ability will grow accordingly.

As designing a fulfilling career is a long shot project, many developers need momentum to get started. Therefore, the last section showed you some fundamental steps that help you gain some quick wins that give you more energy, time, and leverage to make a difference in your career.

Mindset for Career Advancement and Life

The Path Toward Emotional Balance

Many software developers dream about taking control of their careers. Unfortunately, control is just an illusion. All we can do is influence the processes that are often beneath the conscious level.

There is often a clear cause-effect relationship between what you do and what you feel. Most people live their lives in a reactive way: they feel something, then they react to these feelings by taking action. Actions triggered by seemingly random feelings shape their character.

Our goal is to decide on the traits we find valuable, including the development of the four essential soft skills from Chapter 1. Then we take action in the direction of our desired destination and use feelings as feedback.

© Zsolt Nagy 2019
Z. Nagy, *Soft Skills to Advance Your Developer Career*,
https://doi.org/10.1007/978-1-4842-5092-1_2

This is how we gain influence on our lives. This influence is cultivated by emotional balance. As a result of some cause and effect relationships, you can find out the root cause behind seemingly unexplainable states like procrastination or becoming a workaholic. You will get a chance to experience lasting happiness and fulfillment in your professional career and your life. You get a chance to free yourself from status anxiety and social pressure.

We will then understand the importance of self-esteem and self-image and use it to make seemingly hard tasks easier to execute. Swift execution comes from giving yourself permission to do things as a result of respecting yourself more.

You will learn how to use some important concepts such as fixed mindset, growth mindset, comfort zone, learning zone, and panic zone to train and develop your skills. Some examples will help you understand that real growth often happens during the recovery period after training.

We will use some tools to assess and eliminate limiting beliefs that distort our self-image and lower our self-esteem. Oftentimes, some limiting beliefs are formed to protect yourself from deep-seated fears. We will discover how to recognize and understand fears that limit us and address the underlying root cause instead of denying or rationalizing them.

Finally, discovering the motivation trap helps you address the root cause of why many people seem to be unable to intrinsically motivate themselves. Interestingly enough, the real problem is not about motivation, but about another limited resource. By managing this resource and automating a lot of conscious effort, it can be made widely abundant in our lives.

Once in a leadership training, the single most important idea I heard is that you can only help others if you have already helped yourself. In order to take people to the promised land, you have to take yourself there first.

In order to earn more, you have to help your organization. Your contribution is at its peak, when everything is all right with you. Help yourself, before helping others.

Reclaim Your Emotional Balance to Face Career Challenges

Goals of this section:

- Understand how to avoid becoming a procrastinator or a workaholic.
- Discover the obstacles standing between you and a fulfilling career.
- Understand the difference between temporary and lasting happiness.

Happiness and emotional balance influence your career. Balanced people get more things done in higher quality. If you are lacking emotional balance, most of career strategies may not work for you. Yet, with your emotions on your side, you will have an easier time achieving your career goals with less effort. In this section, we will examine what we can do as individuals to preserve our well-being and stay emotionally balanced.

Let's start with two common stereotypes of people out of balance: the procrastinator and the workaholic.

Productivity and self-help gurus often tell us that procrastination is the ultimate evil. Yet, some people procrastinate day and night, and they appear happy. Procrastination is not your enemy. Procrastination is just a way to meet some of your human needs such as comfort. By meeting the same needs in a more empowering way, you can end procrastination.

The real question is: What makes you happy? If you have meaningful goals and you develop momentum, procrastination will disappear. If you just want to earn more money, with the same or less effort, no productivity systems or magic pills will help you overcome procrastination. Define the goals you want, and actions will follow your desire.

Other people are workaholics. They do everything they potentially can to advance their careers, and sacrifice everything else. Workaholics work very hard, sometimes at the expense of working smart. One precondition for working smart is to have a lot of energy. This is exactly what workaholics hardly ever have.

Sacrificing your private life in exchange for getting more things done is rarely useful in the long run.

Some consulting companies force 60- to 80-hour-long workweeks on their employees on a regular basis. The result is obvious: burnout and low productivity.

Working smart requires you to have a work-life balance. Even though work and life are often related, balance means that you are just as happy waking up on Monday as on Saturday. If you can't wait to get home or can't wait to leave home, something is wrong in your life.

Note   According to Jeff Bezos, the founder of Amazon, "work-life harmony" is a better term than "work-life balance" because balance tends to imply a strict tradeoff. We will stick to balance. In my opinion, balance does not need to be maintained with conscious effort. See gravity, for instance. We are in close to perfect balance for life to happen. The heart rate of most humans is balanced, fast enough to stay alive and not too fast to burn unnecessary resources. We want to achieve the same with our lives.

From the point of view of this book, our definition of working smart is to realize our potential in an effective way, while enjoying the process. If this requires competition, compete with your past self and not with others.

In an earlier draft of this book, I called this process "becoming your best self." I now consciously choose to omit this phrase, because it is easy to misinterpret. Imagine if you have to "become" your best self, it implies that right now, your best self is not in you. In reality, this is not true.

In my coaching practice, some of my clients think their lives are broken, because they have not met their goals and they are not as successful as their role models are. They artificially discipline themselves on a regular basis to achieve. Being hard on yourself for completing a project once in a while creates results. However, being hard on yourself all day long in many areas of your life is really horrible in my opinion. It exhausts you mentally. Meaningless hustling is a perfect recipe for a miserable life. Because no matter how successful you are, I can show you someone who is more successful than you are in one area of your life. Think about it, is it worth spending your life disciplining yourself to compete with others?

To lead a more balanced lifestyle, you can also build your framework for success by remembering two important things:

1. You are worthy by just being. You are already good enough. You may not be worthy of a million-dollar salary, but it doesn't mean that you are not worthy as a person.

2. Reaching your goals does not make you happy. Progress causes happiness. As long as you are on your path, you will experience the best time of your life.

It is worth concluding that your "best self" is already in you. It is just hidden by stored trauma and defense mechanisms that surround you. Once you learn how to get past those barriers, you will get access to resources you were not even aware of. As long as you don't have access to your most resourceful states, any tactics may easily backfire on you.

Think about your career as a marathon. There are lots of contestants at the start line. The workaholics start running fast, but take a hit before they are able to finish the race. This hit can be a heart attack, burnout, or lack of motivation. The procrastinators may start late, and lag so far behind, that they may never catch up. The smart workers run at a steady pace, and near the end of the race, they don't face much competition.

It is surprisingly clear why some hardworking people never get ahead: they don't have time to help themselves by balancing their lives out. Smart working people, on the other hand, complete their challenges with ease and stay happy during the process.

The Three Obstacles to Success

If the key to a successful career and fulfilling life is balance, why doesn't everyone do this? I will refer to these obstacles as the 3S list:

1. **Slacking**, that is, keeping the status quo has become too easy, while improvement requires effort

2. **Social pressure**, that is, the opinion of others affecting who you think you are

3. **Status anxiety**, that is, continuously comparing your status with the status of others

Let's examine these three points one by one.

Slacking

Two roads diverged in a wood, and I –

I took the one less traveled by,

And that has made all the difference.

—Robert Frost, "The Road Not Taken" (1916)

In order to take action on your goals on a daily basis, you have to find meaningful goals and think in the long run. In Tony Robbins' first book, *Unlimited Power*,[1] the two twin forces that shape our lives are the fear of avoiding pain and the desire to gain pleasure. For most people, pain often turns out to be a stronger motivator than pleasure.

The reason for procrastination arises from our fear of avoiding pain and our desire to gain pleasure. Taking action right now means pain. The outcome may be pleasurable in the long run, but our minds don't realize it. Not taking action means less short-term pain; and it also gives you instant pleasure by doing something you know well and meeting your needs by checking Facebook, playing a game, or killing time in some other way. The long-term pain of missing out on the benefits is something we barely consider. No wonder why there are so many books on combating procrastination. Whether they help or not is another question.

---

[1] Anthony Robbins, *Unlimited Power: The New Science of Personal Achievement* (Simon & Schuster, 1986).

Another aspect of the pain-pleasure principle coincides with the needs of the Y-generation: instant gratification. People tend to choose short-term pleasure at the expense of long-term pain. Eating a bar of chocolate and gaining weight, having a fun night at the bar but waking up with a hangover the next morning, and so on are all ways of seeking short-term pleasure without considering the consequences.

Every little choice you make shapes your destiny. All little choices add up. Five years from now, you will be very grateful for your present self if you started taking steps in the direction of what you want in the long run.

We are living in the era of instant gratification. Common problems often have accessible solutions. On one hand, life becomes very comfortable this way. On the other hand, following advice of others comes at a price: your own problem-solving skills and your own ability to take action and produce results will diminish because of disuse.

Liking your job may also become far too comfortable. Some people work on a meaningless project at work or select meaningless side-project ideas at home.

For instance, I once interviewed a software developer who was a natural rebel and did everything in his power to avoid contributing to tasks that generated revenue for his company. He used his skills to reinvent a less mature version of popular open source libraries and components. As he took pride in developing and not marketing, he didn't secure sufficient open source support for his projects to survive. He soon searched for meaning at work and concluded that he had to leave his company, because they didn't back him up.

He then continued working on his open source projects without support and was looking for a new job. On the surface, he seemed to be a great hire. He was technologically proficient, and the only reason why I rejected him was that I got in touch with his previous employer and found out about these tendencies after disclosing my fears with them.

On some level, people unconsciously sense when they waste time. Therefore, I suggest working on meaningful side-projects not only to build your portfolio but also to build your self-esteem.

There is a large price tag on comfort. Just knowing how to develop software will give you a good enough salary to live comfortably. No negotiation skills, risk taking, and hardly any social skills are needed for a comfortable job. Career advancement requires all of these skills on some level. If you want to advance your career, but you are afraid of using these skills, you tend to stay in your comfort zone. Your self-esteem will eventually suffer. Your words will become more empty. You will eventually need conscious effort to keep yourself motivated. If this is somewhat true in your life, don't worry. I will show you the way out.

Social Pressure: Are You Living Your Dream or Someone Else's Dream?

People don't tend to take risks, when they are afraid of being judged. Have you ever talked to anyone who chose their career based on what their parents wanted them to become? If yes, you can observe that some people consider the opinion of others more important than their own happiness and well-being. It makes perfect sense to help people you love, especially when someone needs your help. However, giving up on your dreams so that you don't disturb others does not help anyone, and no one needs this sacrifice.

If you want to read a good entertaining story about healthy risk taking and fighting social pressure, I can highly recommend the book *The Third Door: The Wild Quest to Uncover How the World's Most Successful People Launched Their Careers* by Alex Banayan.[2] Alex went to college to study for a respectable job with high income to satisfy the expectations of his parents. However, he chose to pursue his dream to interview the most successful people to find out their secrets to launching their careers. Among other amazing feats, he managed to spend time with Bill Gates and hacked Warren Buffett's annual meeting to ask multiple interview questions. He funded his career by reverse engineering a strategy for winning at *The Price Is Right*. There was nothing ordinary in this story. The reason why I recommend reading it is twofold. First, you will see that a lot more things are possible than you originally thought. Second, even though he faced a lot of anxiety, he was still able to set himself free of it and achieve something interesting.

I also encourage you to free yourself from expectations of others. Find out what you really want, and make changes in your life. You will have a lot easier time with negotiations, if you confidently know what you are capable of.

This step alone will make you think that your opinion is more valuable than before. You will be able to give better advice and create more value. Being the "yes man" or the "yes woman" rarely yields anything of value.

Doing exactly what other people expect you to do is a form of manipulation. The duality of doing your best when it matters and neglecting everything else when you are not in the spotlight eventually taxes your career advancement. This book is all about lasting change. We are targeting a positive, fruitful career path, instead of highly manipulative negotiation techniques, inflating your perceived value to heights you have never reached.

Think about it this way. People worthy of impressing can see through your attempts of impressing them. This leads us to the third S in the list of obstacles.

---

[2] Currency, 2018.

Status Anxiety: Comparing Yourself to Others

Idolizing others is dangerous. Find your own strengths, and improve them on a daily basis. You will find out that in a matter of years, most of your competition will eliminate themselves by not being able to keep up with you.

If you compare yourself to your colleagues and you are only happy if you become better than everyone else, you are destined to feel bad for the rest of your life.

More than a decade ago, I had a very unhealthy mindset at the university. I always compared myself to other students. I continuously beat myself up to achieving on exams and trying to appear knowledgeable instead of becoming knowledgeable. I perfected a system to hack my way through the university exams, but I was lacking real knowledge at most subjects.

I was interested in software engineering and programming. However, I couldn't care less about hardware, electronics, and computer networks. Therefore, after graduating, I needed help with my computer issues and also with basic system administration. I was deeply ashamed of not knowing everything as odd as it may sound. It took me a lot of time until I opened up and started asking for help to get better at DevOps and system administration.

My unconscious decision of trying to appear good in the eyes of others blocked my progress for many years. All I needed was to admit that I needed help.

Instead of asking "Am I good enough?" the right question to ask is "How can I add the most value?" If you focus on value, you will admit to yourself that to give value, sometimes you may need help.

Your career might also be similar in a sense that you are trying to get ahead of people, instead of cooperating with them. See how you can help your perceived competition, and observe your feelings and progress afterward. Notice that people you are competing against now will be nowhere in your life in 5 to 10 years. If you compare yourself only to your past self, you will be grateful one day, as you will develop your skills on a daily basis.

If you think about why you "deserve" to be better than your peers, your career progress will slow down and you will poison yourself with your thoughts. This poison will distort your perception and you will start living in an illusion instead of objective reality. You may state that you have great communication and listening skills while you angrily shout at others and don't even hear what the other person is saying. When you get constructive criticism to improve your performance, you may shut down and use every rationalization technique in the world to justify why the other person is wrong and why you are perfect. This path is called narcissism.

If you choose the distorted path of narcissism, deep down you will still know that you are not perfect, but you may not admit it even to yourself. This creates a self-poisoning effect and suffering, because your perception is already altered, and you will not know why others don't recognize your evident superiority. As your perception is distorted, you won't see that everyone else knows precisely what is going on with you, and some of your peers may even tell you the truth. You are just not ready to hear these words, as you are busy looking at your distorted mirror image you want to become.

Choosing the path of narcissism, your communication and cooperation skills will suffer, and you vastly decrease your chances of getting promoted, or even getting a pay raise. Narcissistic behavior may be a choice from time to time. Once this path is created and enforced, the choice becomes a disorder; and according to the current point of view of psychology, narcissistic personality disorder is not always curable.

Accept everyone else as they are. I discourage you to use other peoples' salary or benefits in your reasoning why you "deserve" more money. You deserve nothing more and nothing less than what you negotiate with your employers.

The perception of status and comparing yourself to others creates performance anxiety. Alain de Botton calls this phenomenon status anxiety both in his book[3] and in his documentary video.

Status anxiety is everywhere. The concept of self-help is defined by status anxiety. Because when we think we are not enough, we suffer. Therefore, it is evident to think that we need "self-help" to help ourselves ease the pain. A new resume or some negotiation techniques may help increase our status, so why not take advantage of them?

As I pointed out before, some of these tools are too hard to use without maturity, because those who are worthy of impressing can see through your attempts of impressing them. This is only the smaller problem though. The bigger problem is if you succeed. Because these tools may distort your perception even further and create a false illusion that everything is all right. With higher status and more power, you are capable of causing even more damage.

Getting a position or a salary with higher status is not a problem as long as it is not pursued for status alone. As long as you focus on increasing your competence, you will naturally be better off, and you will make better use of negotiation tactics too, without suffering from any distortions.

---

[3] Alain de Botton, *Status Anxiety* (Penguin, 2005).

Choosing your attitude will pay dividends in the long run. Your current colleagues may be your business partners one day. Your team lead may provide you with a good testimonial in the future. Above all, it feels good to act with a positive intent.

If I could summarize a strategy for combating the status obstacle, it would be "Do good things instead of worrying about whether you appear to be doing good things or not."

Why Do We Need Balance?

Every system strives for balance. For humans, being out of balance means some sort of negative stress. This stress makes us lose touch with the moment, and therefore, we become unhappy.

The US Declaration of Independence protects "Life, Liberty, and the pursuit of Happiness" and defines this statement as truths to be self-evident. In reality though, happiness is not that easy to define.

In fact, happiness is very subjective. I won't even try to guess what makes you happy. However, research shows that some forms of happiness or well-being are longer lasting.

Hedonic happiness, first described by the Greek philosopher Aristippus, is about gaining as much pleasure as possible and avoiding pain as much as possible.

Note   While Greek philosophers describe hedonism as our fundamental moral obligation to maximize pleasure or happiness, in the ancient Greek context, pleasure or happiness was more of an abstract intellectual process that happens in the mind than what we mean by living a hedonistic lifestyle in a modern world. In the context of this book, we refer to the modern meaning of hedonistic lifestyle, making sure that we are happy by doing pleasurable things.

Notice that instant gratification and procrastination both target hedonic happiness. Taking drugs or alcohol, for instance, also causes hedonic happiness.

Think about the advertising industry for a moment. When you see happy people with perfect makeup in perfect shape, enjoying the product they want you to buy, you instantly feel bad. You might think, "I am not good-looking enough, strong enough, wise enough"; and out of frustration, you go out and buy the product you saw, because you believe the product will make you happy. For a moment, you are the happy owner of an illusion. Yet the illusion disappears right after the purchase.

When you get the raise you wanted for so long, you feel happy and excited. However, within some days, you get used to the fact that you got the raise, you get back to your life, and you will soon start feeling underpaid again. This will make you work on securing the next raise or perk in your life without realizing why you are pursuing securing more resources.

In general, hedonic happiness is rarely a lasting experience.

Aristotle describes an alternative: **eudaimonic happiness**. Instead of meeting your essential needs, eudaimonic happiness concentrates on creating a fulfilling life via growth and contribution.

While hedonic happiness is the happiness of the senses, eudaimonic happiness is the happiness of the mind.

You may now ask: What does eudaimonic happiness have to do with your salary? The answer is simple: you are able to create a lot more value if you grow and contribute. This value will sooner or later manifest in your paycheck.

In the long run, people are looking for value. If you provide this value, others are likely to reward you. Life is about giving and taking. If you have nothing to give, the outcome will hardly be lucrative. So the question is: When can you give the most? It makes sense to conclude that prioritizing your growth and enriching the lives of others ensure that you also create the most value for others.

Exercises

Exercises help you summarize the main points of this book and make you think about your own situation. I cannot stress the importance of these questions enough in your future. You can read all the advice of the world, but if you don't apply it, you will only become more familiar with a dream that never comes true.

I encourage you to keep a journal and write down your answers to these exercises. Self-reflection is one of the most powerful tools in creating new neuro-pathways in your brain. You will start thinking about your situation differently. If you put in the work, you will make the most out of this book.

1. What do working hard and working smart mean to you in your field?

2. Do you have a tendency to procrastinate on actions that are important to you?

3. Do you have a tendency to be a workaholic? Name three ways of working smarter rather than harder.

4.   Describe who you are in a world where you have reached everything you wanted! What skills would you possess? What would your life be like? Where would you be? Whose lives would you affect?

5.   Name three situations at the workplace where expectations of other people influenced you. What difference would it have made in your life if you had stuck to your own path?

6.   Recall your emotions when you compared yourself to one of your colleagues or classmates. What emotions did it create in you? How did it affect your short-term and long-term performance?

7.   Name three means of instant gratification that are harmful for you in the long run. Bonus: In the next 3 months, concentrate on one harmful habit a month and replace this habit with something harmless. For this, you will need to dig deep and find out the positive intent behind you pursuing this instant gratification. What needs are you meeting with it? What else could meet the same needs?

8.   Identify new ways for you to grow and contribute in a professional environment!

9.   What one thing are you going to do today to increase your chances to grow, contribute, and advance in your career?

The Unstoppable Developer

Goals of this section:

- Understand how raising your self-esteem helps your career instead of hindering it.

- Learn a sustainable way of growing without burning yourself.

- Find ways to develop self-esteem without external reference points.

In the first section of this chapter, we talked about the importance of emotional balance and three factors that can easily take you out of balance. We will now continue with a layer beneath conscious action that supports keeping your balance even in situations when things become tough: the layer of self-esteem.

The Value of Self-Esteem

Self-esteem is the act of respecting yourself. Opposed to some self-help advice, self-esteem is not about repeating "I love myself, I love myself," and it has not much to do with positive thinking either.

I have heard so many stories of people being on fire after attending a self-help seminar. They keep on saying "Yes, I can do it!" Then they go out and live the same life as before.

Viewing motivational videos, affirmations, and visualization have very little to do with real self-esteem. Professional athletes do need visualization, because it complements their training. If you don't work hard, visualization only amplifies your dreams.

Self-esteem is the immune system of your mind against stimulations of the outside world. People with very high self-esteem are virtually unstoppable. People with low self-esteem, on the other hand, are very easy to influence.

Modern society comes with instant gratification, smartphone notifications, advertisements, and a high level of comfort. Whenever you put off necessary and important actions in favor of entertainment and comfort, you respect yourself a bit less each time. Your **self-image** will include the description "I cannot solve my problems."

Whenever you feel your self-esteem is not high enough, ask yourself the question: "If I fully respected myself for my actions, what would I do **now**?" A conscious decision on becoming action-oriented will boost your self-esteem right away. As you collect more experience, you will develop judgment that will guide you.

Your self-image is often distorted by your own emotions. Today's world encourages you to choose a very easy and comfortable life. Whenever you find a problem, in the world of instant gratification, many sources encourage you to take a magic pill, claiming that most of the problems in life have been solved by someone.

The problem with magic pills is that they are not effective on their own. Tutorials are excellent tools for shifting your focus and giving you a roadmap. However, no tutorial in the world will solve a problem instead of you. The path leading toward the solution of a problem provides you with more experience and benefits than getting access to the solution.

A strong self-image requires that you command your emotions. This involves conditioning your mind to seek for long-term pleasure, even at the cost of short-term pain. While others are taking their frequent coffee breaks, you may be working on a problem your manager is worrying about. As a result, another time your peers take a coffee break, you may be busy negotiating a higher salary for yourself, based on your results.

I have witnessed extraordinary salary raises, rewarding extraordinary effort, not only in startups but also in corporations. Ben, my client, secured himself a 35% raise, after just five sessions spread over about 2 months. He didn't believe that he had a chance to apply internally for an expert position with a lot more responsibility, even though he was a major contributor to multiple open source libraries. He was coding for fun, and he had a lot of architectural knowledge. Yet, he earned money in the lowest 50 percentile in his position.

Ben thought of himself as someone who could not earn more than what he thought was an average salary. I asked him if his manager knew what he was doing at work on a daily basis. It turned out Ben never even got feedback from his manager.

I asked him what frustrated him the most at work. He said he was often under pressure delivering, as he got the hardest tasks. He also had a lot of pull request reviews, as others didn't stick to the same quality standards as he did.

I asked Ben if he made an effort to mentor his peers. I thought this would be an opportunity for him to act as a bar-raiser, similar to how I improved the interviewing process at the companies I was with. To my surprise, Ben was already a better mentor than anyone else.

Yet, Ben didn't believe that higher salaries were reachable for him. He thought he was just doing what he was expected to do. It took me multiple sessions to access Ben's unconscious mind when he finally admitted that he thought he did not deserve more money, because making more money was dangerous. During his childhood, Ben witnessed his father put himself into a really tough spot. His father assessed a business risk wrongly. Instead of making a lot of money, his father got into debt, and this debt caused a lot of trouble for his family. Ben's father shifted back and forth between frustration and anger, and Ben hardly had the money to buy the mandatory books for school. Ben unconsciously concluded from the events that aspiring to make more money than what you are earning right now is dangerous. You may lose everything. Taking risks is dangerous.

Ben made an unconscious decision that pursuing a lot of money was bad for him. As he also had to be a top achiever to mildly uplift the mood of his father giving him some hope that one day things would change, this achiever mentality helped him develop a razor-sharp problem-solving mindset. Therefore, he always overperformed, while he was underpaid.

After helping Ben realize that he had more value than he thought, I spent a session with him to prove that he could ace any interview and get any job he wanted. I gave him a tougher coding challenge than what I asked in an interview. He nailed it. I also collected evidence from HackerRank how tough this challenge was. He thought it was a simple task.

Ben still struggled believing that he could get a job. So I had to prove him that getting a job is the norm for even developers with below-average skills. I showed him stats about unemployment rate in IT engineering. The rate was so low that he slowly but surely realized that odds were in his favor to get a job.

After all that, he was still afraid of communicating with his boss.

So I introduced him to the book *The Six Pillars of Self-Esteem* by Nathaniel Branden.[4] In the book, the author defined what you need to work on when building your self-esteem. These pillars are as follows:

- **Living consciously**: Know your goals, values, and beliefs and act accordingly. Working on your consciousness is required for controlling your emotions. Our brain works on autopilot most of the time. If you feel bad about not earning enough, your feelings will manifest in worse-quality work, decreasing your chances at the negotiation table. If you feel you are not good enough, you will not claim enough value in return. Pay attention to your feelings, and autopilot actions. Get rid of patterns that stand in your way.

- **Self-acceptance**: Value yourself, treat yourself with respect, accept your mistakes, and forgive yourself. For any self-improvement to take place, it is a necessary condition to accept yourself. You don't have to love the way things are working out for you, but don't be your own enemy either. You don't have to like your mistakes, but you should be able to forgive yourself.

- **Self-responsibility**: You are responsible for your choices or actions. Taking responsibility comes with tension. It is always easier to rationalize your situation and blame someone else. Making a victim out of yourself is harmful for your self-esteem. You are in control of your life, and each time things don't work out in the way you want them to, always think about the factors you are capable of influencing.

---

[4] Nathaniel Branden, *The Six Pillars of Self-Esteem: The Definitive Work on Self-Esteem by the Leading Pioneer in the Field*, reprint edition (Bantam, 1995).

- **Self-assertiveness**: Stand up for yourself openly, honoring your needs and values. Never fake who you are for fame, money, or other gains. In order to practice self-assertiveness, you have to accept that your thoughts and desires are important. Whenever you sense that you tend to avoid conflicts or try to meet someone else's expectations instead of your own, you are on the wrong track. Whenever you stand up for yourself, your self-assertiveness grows.

- **Living purposefully**: Your goals are important to you, and you take action toward reaching them on a consistent basis. These actions are also productive. If you know your outcome, you know what you are moving toward. This determines how much money you are targeting at the negotiation table. Productivity is meaningless on its own. Productivity should only be targeted, once you know your purpose. Procrastinating actions should be identified. Your self-esteem grows when you eliminate procrastination.

- **Personal integrity**: Honesty, reliability, and trustworthiness. Your behavior should be congruent with your values. Whenever you know you are lying, your personal integrity is at stake. Whenever you are not self-assertive, you are not acting with integrity. Whenever you procrastinate, your actions are not in alignment with your goals.

Which of these pillars were shaking in Ben's example?

He was unaware of the effects of his beliefs about money, so he was sabotaging himself. His past trauma, his behavior at work, and the simulated interview all enforced that he had a subconscious block to overcome.

Ben didn't have many problems with self-acceptance. As soon as he found out about the problems his beliefs caused him, he started taking action. He was just afraid of taking the first steps. Fortunately for him, another two pillars, self-responsibility and personal integrity, helped him a lot.

Initially, when I asked him about his purpose, I got some uncomfortable pauses. He got some sense of wanting to become a great professional. Based on what he told me, chances were he was already a great professional. So we worked a bit on discovering his purpose. You will find some tools for this discovery in the next chapter.

The biggest problem on his end was assertiveness. To be assertive means you stand up for your own interest, while considering the interest of others. A submissive person only considers the interest of others. An aggressive person only considers his or her own interest. Ben was very much on the submissive side thanks to trauma coming from his past.

While I am not in a position to heal trauma on my own, I could point him to resources. One thing that would help him the most in the shortest amount of time is assertiveness training. While he could research training in this area on his own, I pointed him to the book *When I Say No, I Feel Guilty* by Manuel J. Smith.[5] This book details a lot of great stories on how to be assertive.

When it comes to workplace assertiveness, *The Clean Coder* by Robert C. Martin[6] helps you a lot, because it uses stories of a software developer not only to illustrate the principles of being assertive but also define the boundaries of a software professional.

These two books inspire you on when to say yes, when to say no, and how to stand up for yourself even if the situation becomes uncomfortable. We discussed some other long-term plans that you will find out after the end of the success story.

During the fifth session with Ben, we worked on a positive trait of assertiveness. I disclosed my experience that talent is sometimes rewarded without negotiation. If a junior employee performs on senior level, their salary will often catch up to senior level, without extra negotiation.

However, this reward is rarely the maximum the candidate can get. Negotiation creates even more value.

The problem iss deep down Ben still believed that by negotiating, he takes value that he does not deserve. This was when I put him off balance with the statement that I believe it iss his duty to negotiate, because I strongly believe that by claiming more value, he creates even more value for his employer.

Negotiation is not about distribution of limited resources. It is rather about making the whole pie larger for everyone. This is why FBI hostage negotiation does not apply to getting raises. In hostage negotiations, the criminal walks away with nothing while the negotiator agent prevents a hostile situation. The pie was not meant to be distributed.

---

[5] Manual J. Smith, *When I Say No, I Feel Guilty: How to Cope, Using the Skills of Systematic Assertive Therapy* (Dial Books for Young Readers, 1975).
[6] Robert C. Martin, *The Clean Coder: A Code of Conduct for Professional Programmers* (Prentice Hall, 2011).

> **Note** You may be aware that I am using FBI hostage negotiations as an example because of the excellent book of the former FBI agent, Chris Voss, titled *Never Split the Difference: Negotiating as if Your Life Depended on It.*[7] I highly recommend not only reading the book but also watching any videos you can find with Chris. I recommend looking into Chris' work because I strongly believe in tactical empathy.
>
> However, apply the principles of this book with judgment, because your situation may be different. Suppose you are negotiating about the distribution of a pie. In hostage negotiation, the agent takes the whole pie, leaving nothing for the other party. In software development, your optimal negotiation tactic is to go for drastically enlarging the pie. Even though the principles of *Never Split the Difference* work perfectly while enlarging the pie, many authors focus on taking the largest slice of the pie instead of doing anything to enlarge it. Some authors in the area of salary negotiation paint the company as the enemy, and they spread advice that makes you focus solely on taking. This approach backfires in environments where cooperation is rewarded more than competition.

At a working environment, the employee lives in a symbiotic relationship with the employer. The employee can earn less on their own, while the company cannot function without employees. If the employee creates more value, there is a leverage effect through the resources the company can provide.

Salespeople know this really well. If you negotiate well and close a $10,000 deal and you get 10% commission and you speak to 100 people a month, you get a chance to create up to a million dollars of value. By increasing your closing rate by just 1%, you create $10,000 of value. $9,000 goes to the company, and $1,000 goes to the salesperson.

Software developers often add just as much value as salespeople do. Unfortunately, this job neither is tangible nor requires the developer to negotiate on a regular basis.

In Ben's case, he was recruiting for a position requiring more expertise and mentoring skills. Ben was too shy to consider himself senior enough to apply. I pointed out that he was already doing most of the responsibilities the role required. However, he was expected to do a lot more lower-level stuff that less senior members of his team could also do, especially if they were mentored by him. Therefore, taking the new position creates a win-win situation for everyone.

---

[7] Chris Voss, *Never Split the Difference: Negotiating as if Your Life Depended on It* (Harper Business, 2016).

When negotiating, you have to concentrate on how to make the pie larger. One way to make the pie larger is to be transparent about what you are already doing and what you are planning to do in the future. This is why it makes sense to send a weekly report to your lead, especially if you are not yet targeting a raise in the short run. Are you mentoring your peers? Ask for this responsibility in a structured way and tell your lead what outcome you expect from the sessions. This will make you more goal-oriented, so everyone wins.

Is a process missing? Help your team introduce it. Are you lacking knowledge? Ask for expensing a certification training and get certified fast. You cannot help your company without negotiating.

This is why I believe people with extraordinary talent have a duty to learn how to negotiate. First of all, not all companies reward extraordinary talent. Second, negotiations may still create value for both parties.

High self-esteem is key in executing any salary negotiation strategy effectively. If your self-esteem is low, you may not even think you deserve the space you are occupying. You may think that your voice should not be heard. You may think that your opinion is not important. Good luck convincing your boss or your HR manager that your contribution is worth a lot more than what you are currently getting.

Ben started getting it. If he wants to be the best professional, he has to enlarge the pie. To enlarge the pie, he has to have self-esteem. He can only have proper self-esteem if he charges money for his services, creating a win-win situation.

Unfortunately, many people panic in this situation. So I asked Ben for a role play. I played the role of his line manager, and he had to convince me to give him a chance. First, he crashed and burned, because after a few sentences, he stopped believing in the value he created.

You might have heard of the **comfort zone**. When you keep doing things that are comfortable for you, you stay in your comfort zone. As soon as you start growing, you leave your comfort zone. Growth is achieved in the **learning zone** around your comfort zone. Stretching may make your comfort zone permanently larger, as you consistently face your fears.

Your growth should not be too abrupt, as you may end up getting into the **panic zone**. When people panic, they return back to their comfort zone, with some side effects they need to cope with later.

Ben's freeze in the role play was a great learning zone experience, because it was uncomfortable enough for growth, but Ben was intelligent enough to know that it had no negative consequences. We went through the material once more. He finished it for the first time. His tonality still projected that he didn't believe in his message fully, but at least he executed it. I asked him to practice it more.

Two weeks later, Ben reported that he got a 35% raise. At first, I was puzzled, because I didn't expect this result so fast. It turned out that on the night of the previous session, he constructed an irresistible offer, detailing what he was doing at the company and what he was planning. He requested a meeting from his manager the first time in his life. He worked on his assertiveness before the meeting, because there was no way to cancel it anymore.

Ben and his manager went through Ben's plan of increasing his contribution. He proved his competence and plans during the meeting to function well in that higher position. Ben also shared his frustrations about the PayScale data he had collected, and he also disclosed that he was happy to work with his manager to close the gap according to the requirements of his new position.

Credit also goes to Ben's manager, because he recognized Ben's skills and attitude. A 35% raise in a corporate environment is by far not a typical result. However, the internal transfer secured him the budget, and it was evident that a change had to happen.

Ben also reported that he got a chance for another 20% raise on top if he fulfills his expectations. This is because the 35% raise resulted in an average pay for his new position. But both Ben and his manager knew that he could do better. If you do the math, 1.35 * 1.2 = 1.65. Ben got a chance to raise his salary by up to 65% if he meets some goals.

As we still had some sessions left, we worked hard on documenting Ben's journey to make sure the 20% raise would also happen for him as fast as possible. The yearly salary review cycle was due within 7 months, so we targeted this time span for his second raise.

Was it evident for you that Ben deserved these raises? It was for me. Unfortunately, Ben didn't believe that he deserved these raises without a lot of extra work, because he felt guilty. This is why raising your self-esteem is important.

Approaches for Building Self-Esteem

The six pillars of self-esteem describe the areas of your life to work on. Unfortunately, knowing these pillars is not enough. If knowledge was power, we could just read a book and become unstoppable.

There is a reason why the books *Unlimited Power* and *Awaken the Giant Within* by Tony Robbins are fairly inexpensive, yet, Tony can charge several thousand dollars for his seminars for tens of thousands of people every year. Most of the information Tony teaches is in his books. Many people who buy his book never open it and read it. Many people who buy and read his book may procrastinate on the action items of the book, and they eventually forget taking any action on it. For most people who buy these books, the desired transformation never happens, because not everyone can put advice into action without action items and additional encouragement.

This is why I decided on putting exercises at the end of most sections in this book. They are not there for you to read and skip. They are not there so that you deliver minimum effort and tick a box that you completed them. They are there to change your life.

Regarding building the six pillars of self-esteem, you need to take action if any of the pillars are weak in your life. Depending on the pillar, there is an order at which work should move ahead. For instance, if you are lacking self-acceptance, there is no reason to target any improvement before you accept yourself as you are.

Actionable advice on building your self-esteem is not easy, because there is no one uniform path that helps everyone.

One universal truth about building self-esteem is that there needs to be a synergy between emotional work and real-life action.

- **Emotional work** is an inside-out approach. You work on your emotional intelligence through a therapeutic and/or coaching process that results in an emotional transformation. This transformation enables you to give yourself permission to act in ways that you used to block or sabotage because of fear or discomfort.

- **Real-life action** is an outside-in approach. You collect **reference experiences** in the real world until you get desensitized to the sensation that you were afraid of. Reference experiences may trigger deep emotional work, but not for everyone, and the process is not controlled.

Outside-in is not enough on its own, because it does not foster emotional transformation. The outside-in approach may give you an illusion of progress, because you get results. However, as soon as you stop the process, your comfort zone shrinks again to a level where it was before.

Note In his book, *The 4-Hour Work Week*,[8] Tim Ferriss introduced the comfort zone challenges such as lying down in a busy place. Many confidence coaches bought into the idea, fueled by the rationalization that it is a form of cognitive behavioral therapy (CBT). This is a perfect way to create the illusion of success. Comfort zone challenges rarely trigger emotional transformation. As soon you stop doing the challenges, your comfort zone will start shrinking again.

---

[8] Timothy Ferriss, *The 4-Hour Work Week: Escape 9–5, Live Anywhere, and Join the New Rich* (Vermilion, 2011).

The role of an outside-in process is to give you leverage and quick wins to stay invested in an inside-out emotional transformation process. Outside-in is also great to measure how you are maturing emotionally.

You may find yourself crash and burn at a tough negotiation or public speaking challenge. If you do outside-in, you may develop your skills to an acceptable level. However, once you stop public speaking for years, you may start doubting yourself once you start again from scratch. This is because the underlying core insecurities were never addressed.

At the same time, if you do inside-out work, you may find that your public speaking or negotiation skills greatly increase without any conscious work on them. This is because the skills were always in you. The reason why you couldn't perform is that your core insecurities prevented you from doing your best.

This is why I am prioritizing the inside-out transformation. I will recommend you a couple of steps I tried out myself.

Emotional Work

In the long run, **therapy** is highly beneficial. One great option is to participate in **Internal Family Systems (IFS) therapy**. I participated in multiple sessions myself, and I managed to understand and let go of a lot of baggage I got from my past.

Cognitive behavioral therapy (CBT) helps you with the synergy between emotional work and real-life action. **Assertiveness training** is one form of CBT. CBT replaces ways of living that do not work for you with strategies that do. Then you get a chance to observe and self-reflect on the benefits and drawbacks of your new approach. Therefore, CBT connects real-life action with emotional work. The problem with CBT is that the emotional transformation is not necessarily deep. For instance, you can become assertive without addressing the underlying trauma that prevented you from standing up for yourself, and this trauma may hurt you elsewhere in your life and career.

A deeper form of therapy is **grief work** with a good qualified psychotherapist. Grief work addresses traumatic events in your memory that you may not be consciously aware of. It helps you relive memories that block you from respecting yourself without being an achiever or a pleaser. This work has to be done either by qualified professionals or by people who work with their own emotions for a living.

Quiz time. Which profession requires you to work with your own emotions the most?

The answer is fairly simple – acting. Actors and actresses are the most credible source for a high degree of emotional awareness in my opinion. I chose the type of acting that requires you to get in touch with your own emotions to tap into an emotional state. This form of acting is called **Stanislavski-based method acting**.

Emotional memory can create realistic performance. Method acting requires you to access your own emotions to represent a character. Recall the performances of Heath Ledger as the Joker, or Marlon Brando in his various roles in *The Godfather, On the Waterfront,* and *Truckline Café*; these are great examples for method acting.

As actors work with emotion, visiting a method acting coach or workshop may teach you a lot about how you relate to your emotions in practice.

If you consciously work on your self-esteem, one day your colleagues will notice that you changed before you notice it. Your productivity, willpower, and the weight of your opinion will skyrocket. You will be a lot more comfortable with who you are. This makes it not only easier for you to negotiate your salary, but you will also be the right candidate for a future promotion.

Instead of searching for external validation by getting other offers, you often don't even have to leave your job to get promoted. You may not even have to initiate salary negotiations.

Real-Life Action

Some people, especially those diagnosed with narcissistic personality disorder, disdain emotional transformation. They claim, "Working on your emotions does not make you a better chess player if you don't even know the rules." Unfortunately, this statement is a dangerous half-truth perfectly suited for spreading emotional fog.

I have had the privilege to play a poker tournament with a €1,500 buy-in. I identified a situation where I could more than double my chips, and I estimated my chances to win at around 2:1, meaning that I win twice for each loss. As it was early in the tournament, the optimal play resulted in a profit of more than €600. I had a choice to make:

- Submit to the aggression of the opponent and sacrifice the €600 expected profit in exchange for a 100% certainty of staying in the game.

- Risk getting eliminated one out of three times of the time in exchange for a €600 expected profit and better opportunities later.

I knew the right action. However, as I was reaching for the chips, they felt heavy. My body started shaking. I felt bitterly uncomfortable. Back then, the €1,500 buy-in was a lot of money for me, as I just finished university, and it was more than 2 months of my net starting salary. Eastern European starting salaries were low.

I took a deep breath. Notice breathing often helps. After tossing all my chips in the middle, I felt shame and anxiety. What if my cards were not good enough? How embarrassing would it be to run into a very strong hand?

When I revealed my hand, the whole table got shocked, because they were risk averse themselves and thought I had a lot stronger hand than ace-jack suited. My opponent couldn't believe his eyes, stood up, and started shouting for a ten or an eight, because he happened to put all his chips in with those hands assuming I submit to his aggression just like everyone else at the table did. He caught an eight, and I got eliminated.

Most of the table concluded the lesson that it iss wise to fold an even stronger hand here, because you can get eliminated at any time. My lesson was different though. I concluded if I don't take these calculated risks, I will create an illusion of progress by staying in the game in a worse position, while aggressive players will continue exploiting my weakness. I would rather maximize my chances of winning than setting myself up to lose as late as possible.

I concluded that in order to set myself up to win, I have to become comfortable with the uncomfortable. This requires emotional work. I found sources for emotional work in meditation and sport psychology, and I studied all sources of achieving peak performance.

Years later, I got a chance to negotiate a freelance contract. I got some insider information that $100/hour was a rate I could get away with. Back then, I made around $40,000–$45,000 a year. This means my hourly rate was less than $25 assuming 230 working days and 8 hours a day.

Remembering my poker experience, I told my client my rates were $100 an hour without hesitation. I felt anxiety, but fortunately, the negotiation went over the phone, and I could moderate my tonality. My client quickly said OK and inquired about what else I can help him with. Anxiety quickly became excitement. This became my reference experience for where I became comfortable with the uncomfortable.

Everything adds up in life. Every little action you take creates thoughts and emotions that either build or undermine your self-esteem. Most salary negotiation materials forget that negotiation starts with an inner game.

Higher self-esteem comes with an additional benefit that you believe that you are mature enough to undertake more responsibilities. After all, associating higher value with your efforts is a form of respecting yourself. Your goals will also be higher. If you are earning $60,000 now, with low self-esteem, you may target $65,000, while with higher self-esteem, you might even take a shot at a significantly larger raise. If you aim for $90,000, you won't be disappointed to walk away with $80,000.

When you respect yourself, you will create less excuses. We all know the anxiety of starting a discussion we feel uncomfortable with. Memorizing phrases, learning framing techniques, learning FBI-proof body language, and sticking to a script will hardly help you in the wrong state of mind. It makes a lot more sense to decrease anxiety in you first. The exercises at the end of this section, and the elimination of self-limiting beliefs, will guide you in this process.

If you are afraid of negotiations, start small. The following tasks may help you, if you apply them in the right context:

1. Start greeting everyone in your office, with a downward inflection in your voice each morning you arrive.

2. Ask a person you are not close friends with about their day.

3. Ask the opinion of one of your colleagues about a topic you are interested in.

4. Mentor a less experienced colleague of yours a bit.

5. Ask your lead, manager, or boss about their future plans.

6. After making a mistake, tell one of your colleagues about it, and talk about how to avoid such mistakes in the future.

7. Ask your lead, manager, or boss for 5 minutes of their time and suggest a small, meaningful improvement about the current workflow in your company. Make it natural; make it meaningful.

8. In a meeting, if you didn't fully understand something, openly confess it, and ask for clarification.

All these exercises make you more social. In addition, some of these exercises may even increase your perceived value. In the end, you will find that you are a respected member of your organization.

Regarding point 8, I once worked with a great freelancer. He had to represent us in meetings with business stakeholders. In these meetings, due to the application of business jargon, sometimes none of us knew what they were talking about. As soon as he got a chance to talk, he said, "Excuse me. My English parser is a bit rusty today. Could you please simplify what you said?" This was authentic, because his first language was not English. The whole room, including the businesspeople, started laughing.

Some of the actions I am suggesting may be uncomfortable for you. However, your self-esteem grows in this period, especially if you decide that you will tolerate the pain in exchange for growth. As you make progress, each task is just a bit more difficult than what you are comfortable with.

Some of my readers may be in a harder situation. Suppose people avoid you, they treat you badly, and don't respect you. I understand it is harder to build self-esteem without external acknowledgement. However, notice the word self in self-esteem. It can be built without taking the opinion of others into consideration.

Coaching, Consulting, and Mentoring

Many people start with coaching, mentoring, or consulting, thinking it is a substitute for therapeutic work providing emotional transformation. While a great coach and some excellent mentors may trigger emotional transformation, in reality, coaching, mentoring, or consulting are not meant to substitute work on inside-out transformation. Eliminating your limiting beliefs, understanding your fears and phobias, and working on your unconscious are mostly done in a therapeutic context. In order to understand the differences between coaching, consulting, mentoring, and therapy, I will clarify what expectations you can formulate when using these tools.

Both **consultants** and **mentors** are experts at a certain field. If I want to learn how Google Cloud Platform works, I can obtain mentorship from a certified course instructor or subject matter expert. The difference between the two is that a consultant is hired to solve a specific problem. A mentor is hired to help you reach a larger goal. In this book, I am mostly mentoring you based on my own plan. I can relate to you, because I can understand why you bought this book. Right now, I am mentoring you on tools to use to build your self-esteem. I will also mentor you on personal branding, interviewing, and negotiation, assuming that you find answers to questions you are looking for. In private mentoring sessions, this approach can become a lot more tailored.

A career consultant may specialize in resume writing, personal branding, interviewing, or negotiation. You buy actionable knowledge that you need in your own special situation.

A **coach** may or may not be an expert in the area of expertise of the client. A coach asks clients the right questions so that they find out the answers themselves. A consultant creates a plan for a client that the client has to execute. A coach asks the right questions so that the client can come up with a plan on his or her own.

Coaching, consulting, and mentoring rarely facilitate inside-out transformation on their own. A good coach may trigger you to perform inside-out transformation on your own. However, coaching is not a substitute for therapy.

The closest you can get to an emotional transformation experience with a coach, consultant, or mentor is in the area of the previously mentioned field of method acting. My method acting coach, Shredy Jabarin,[9] leveled me up in storytelling, helped me overcome anxiety and emotional challenges at work, and helped me shield myself from emotional toxicity of people with predatory behavior. Depending on the session, I got coaching, mentoring, or consulting. The side effect of this work has been an emotional transformation that allows me to record live videos, while back in 2015, it took me 2 weeks to record a 10-minute-long screencast, where I didn't even show my face.

Back in 2015, I lacked the self-esteem required to record myself in camera. Today, I just don't care. It does not have to be perfect, I just do my best, and that's enough.

Exercises

1. Practice the eight communication exercises described in this section. If you are afraid of the workplace, go elsewhere, and adjust the situation! Watch yourself grow.

2. Determine where you are now. Describe how the six pillars of self-esteem manifest in your life at the moment. Focus on both positive and negative aspects.

3. Describe how you want the six pillars of self-esteem to support your life. What would you gain by living this life?

Taking Responsibility Is an Inevitable Ingredient for Growth

Goals of this section:

*   Compare the chances of people in fixed mindset and growth mindset.

*   Learn how to use post-traumatic growth to your advantage.

*   Learn how to recognize and eliminate situations when you feel helpless even though you could do something.

---

[9] www.shredyjabarin.com

When things go well, most people take the right actions and move toward their goals. When things go badly, most people give up. This is the reason why I used the marathon metaphor in the first section, to illustrate performance at work. Competition seems to be intense initially, but as time goes by, most people simply don't keep up with the pace. Some drop out after the first obstacle; others procrastinate for years.

Notice that the exact same events may shape different peoples' lives in different ways. Difficulties traumatize some people, while others see the opportunity of massive growth.

Suppose you do everything perfectly and you change everything you possibly can in your company. One day, you initiate a performance review, you ask for a raise, and the stock market crashes. Doing the right thing at the wrong time is very painful. The question is what you do with this situation.

Some people decide on blaming their strategy, their managers, and the outside world. These people are convinced they deserve a lot more than what they are currently getting. They become result-oriented and stop taking action that would have made them successful had the market crash not happened.

The other approach is to take full responsibility for your actions. You can differentiate between things you can influence and things you can't. Short-term results don't matter. In the long run, your actions will add up to a destination.

With the right approach, even if the market crashes, your colleagues will back you up; and the next time the opportunity arises, you maximize the chances of getting a significant raise, even without asking for it. If you leave your current company in favor of growth elsewhere, you will still be backed by a lot of people. Who knows when you are going to work together again. The world is very small after all.

Events that happen to us are mostly neutral. What shapes our lives is our internal monologue. This monologue associates a meaning with events that happen to us. Some interpretations may lead to growth, while others may lead to destruction.

A person with a **fixed mindset** thinks that our abilities are determined right after birth. Some people are gifted with talent, while others are not. Failure is fatal. You feel defeated whenever you find out that you are not good enough.

According to **growth mindset**, talent is nothing else but your current state. Instead of talent, effort matters. Failure is not fatal, but rather essential. You only get defeated, once you stop trying.

More details on these two mindsets can be found in the book *Mindset* by Carol Dweck.[10]

As a sidenote, I am not a fan of preaching that you can reach anything in life. Chances are that you are not going to become an Olympic athlete if you have lived 30 years of your life sitting on the couch and you don't possess genes that enable you to compete on Olympic level. However, great things can still happen if you apply growth mindset and you push yourself. You can compete against your past self and explore your boundaries, giving you a journey with constant growth experience.

The number one difficulty with growth mindset is that you need to grow. It feels very uncomfortable to form meaningful connections with your colleagues, if you were originally an introvert guy sitting in front of a computer all day, writing code. This is why you need to respect the boundaries of your **comfort zone, learning zone**, and **panic zone**.

When you start working out, initially you are weak. You are not going to start lifting huge weights right away, as you risk massive injury. You grow in your learning zone. Use lower weights initially that are just a bit heavier than what you are comfortable with. Then increase load gradually! By delivering just a bit more than before, these baby steps will add up to a sustainable growth path.

Action is very uncomfortable in the learning zone. However, facing discomfort will make you grow while you recover from the temporary loss of comfort. You may not feel this while taking action, but the effects will manifest later. In the gym, your muscles get hurt in the **learning zone**. Your body repairs these muscles such that they become stronger than before. This is the idea behind **post-traumatic growth**.

Note   There is a relationship between your self-image and the size of your comfort zone. The human body tends to stick to a baseline. Imagine it like a thermostat. If things are going badly, it heats things up. If things are going too well, it cools things down. The baseline is determined by your comfort zone. Through working on your self-image and spending more time in the learning zone, you can continuously increase your comfort zone.

One of the reasons why some people never grow is **learned helplessness**. If you believe that salary raises are fixed, they are mapping the inflation rate of money at best, and you cannot do anything about this, you are not encouraged to leave your comfort zone. In practice, you will see some people getting no raises for years. At the same time, other people double their salary. If you are motivated in getting a raise, you will find a way.

---

[10] Carol S. Dweck, *Mindset: The New Psychology of Success*, updated edition (Ballantine, 2007).

During your work, you may feel that some steps you need to take are not too comfortable. Public speaking, meetings, voicing your opinion, or even saying "hi" may be very hard for you.

Learned helplessness is connected with expectations of other people. During childhood, most people are conditioned to meet expectations of their parents and teachers. After our studies, expectations of society take over. Whenever you feel you should do something, but you are afraid of pursuing your dream because of expectations of other people, the result is confusion. There are two paths leading from here: take responsibility for your actions by deciding to make your path or grow learned helplessness by giving up.

In order to grow, you have to take some bold moves. Make sure you stay in your learning zone to achieve growth, and try to avoid the panic zone. Take calculated risks, not suicidal ones.

Growth mindset will not only help you negotiate a higher salary, but it will also increase the value of your work. One day, you will figure out the raise you were asking for is just a fraction of what you are capable of reaching.

Exercises

1. Name three areas in your professional life where you experienced learned helplessness.

2. Plan one action in each area to act against learned helplessness.

3. Do you ever lose control of your emotions during work? Why? What can you do to stop losing control?

4. When do you feel stress during work? Which types of stress are good for you? What can you do to stop feeling stress that is not good for you?

5. Name five areas in your professional life where growth mindset will help you reach your goals faster. Describe your growth process.

Limiting Beliefs in Your Career

Goals of this section:

- Experience what it is like to be in the box and out of the box.

- Learn how to eliminate your limiting beliefs using the ABCDE method.

The quality of your work depends on the quality of your thoughts. An average human has 60 to 70 thousand thoughts a day. If these thoughts tell you that you are weak, clumsy, and ugly, your self-esteem will weaken, and your self-image will be shifted accordingly.

Another aspect of your thoughts is that you are responsible for whether you are **in the box** or **out of the box**. If you are in the box, your limiting beliefs override your rational thoughts. Typically, you say things like

- I can't talk to my boss.
- I shouldn't think about raises.
- I won't be able to negotiate.

There is a completely different world outside the box:

- I choose to give my best every day.
- I want to and I can make positive changes in other peoples' lives.

In the box and out of the box thoughts have not much to do with positive thinking. You are not forcing positive thoughts on yourself. The reason why the messages sound more positive out of the box is that you are less limited there and you have the option of choosing an internal monologue that suits you better than in the box.

If you believe in fixed mindset, others put you in the box. If you believe in growth mindset, you are responsible for putting yourself in the box.

If you accept that you are responsible for putting yourself in different mental states, your self-esteem will eventually grow. Your decisions will be better, and your work will become more valuable.

Most of the time, your internal monologue puts you in the box. People tend to narrate what happens to them. Sometimes this narration becomes very negative. We often distort reality; our conclusions are not logical at all.

Your internal monologue often creates obstacles that hinder you from reaching your goals. Your emotions in the box are usually less healthy, and you can often convince yourself to act against your interest.

For instance, if you cannot accept yourself, the way how other people appear to judge you matters more to you. Of course, everyone appears to be judging you in the box.

Example: Wrong Attitude

Frank gets a job, and his starting salary iss $55,000. He asked for a salary inside the range of $52,000 and $60,000. Following the advice of a salary negotiation site, he wrote an email saying that he would appreciate earning more. However, HR threw the ball back, saying that they had met Frank's expectations and they had even offered him more than what he had asked for.

Before working, Frank went to a New Year's Eve party, where some of his friends, also working in the same industry, told him that he had been lowballed. According to his friends, Frank shouldn't have accepted any offers below $75,000.

From day 1, Frank's attitude and thoughts were doomed. Instead of desperation, Frank tried to find leverage and started working efficiently. He started competing against his colleagues. He often had thoughts like "Damn, this guy must be earning at least $70,000. I am exploited! I will resign as soon as I get a better opportunity."

Frank made sure that his work appeared superior to the contribution of everyone else. Instead of cooperation, he created competition. Each task he accomplished was proof to him that he was better than the others.

A couple of months later, Frank started interviewing with other companies. During the selection process, whenever he had to talk about salaries, his fluent, smooth tone changed to a rigid tone, grasping for breath. He kept thinking that all companies wanted to exploit him.

Eventually, Frank got an offer of $62,500. It was better than $55,000 after all. So he sent an email to the head of his department, asking for a feedback talk. Given his boss was busy, he did not react on the same day, not even noticing how important the message was to Frank.

After all, when Frank wrote the request, he only mentioned a feedback talk. He was far too afraid of writing more in an email; his internal narration told him it was not appropriate and everyone would judge how greedy he was.

The day after, Frank found out that his manager was on holiday for 3 days. He sent his application to five other companies on the same day.

During the feedback talk, Frank showed every little proof about his superiority and listed five things that he thinks the company would need according to his judgment. Then he said he was very disappointed with his salary. He voiced he would deserve a lot more, and he expressed he knew everyone else was earning a lot more than him.

Once the discussion ended, Frank's manager started thinking about the situation. He really liked Frank until the feedback talk. In fact, he was considering putting him on a list for a larger than normal raise. However, after the talk, he lost trust in Frank's attitude.

Given that there was a shortage of employees in Frank's department, he got $64,000. However, Frank had to pay a big price for his negotiation efforts: he was never promoted afterward. Even though Frank had the best skills, he was proven to be very hard to handle.

One day, he found out his colleague got promoted. Frank exploded: "That idiot! He cannot even spell the word promotion!"

Frank started interviewing elsewhere and got a job offer for $65,000. Frank told his boss he got an offer for $75,000 and showed him benchmarks, proving that he was earning less than the average salary for his position. This benchmark was all fake, as he carefully deleted most startup offers, which were significantly lower.

Frank's boss recently had an option of hiring an excellent developer, but he didn't have any reasons to fire anyone from the team. Given the history and Frank's mental state, he concluded that he had to let Frank go. Frank quickly replied to the other company, saying that he would really like to work with them, but actually, his skills would be a better match for $68,000, as he deserves a higher amount for the torture he has to go through with the slightly worse working conditions than what he's used to.

Eventually, the other company withdrew the offer, as they found Frank's email highly unprofessional. Frank lost both of his options.

Example: Right Attitude

Suppose that John also got a $55,000 job and he found out the market was paying $75,000. Once John got started, he tried everything he possibly could at the startup to learn, improve his skills, and create value.

Instead of dealing with tasks that looked nice, he was trying to understand what moved the company forward. He made a lot of suggestions to his manager, some of which they started implementing together.

During one lunch break, John mentioned to his manager that he really liked his job and making a difference matters to him. The only thing bothering him though is his knowledge that he could earn $20,000 more elsewhere. He also revealed that he only found out about this after accepting the offer. John was a bit uncertain, as he did not want to leave the company.

The day after the informal talk, John's boss gave him a contract amendment to raise his salary to $60,000 and a $5,000 end of year bonus. Then he showed John that the benchmark in startups is actually at $60,000, and he could only get the $75,000 salary if he worked with different technologies.

Furthermore, John's boss sat down with him to tell him what he needs to do to earn $75,000. In 8 months, John got promoted, and he earned $75,000 as a tech lead.

Compare the Two Attitudes

Notice that Frank's mindtalk didn't leave him any rest. He must have had a very frustrating career, always looking for the next raise. Each and every day, he found a way to feel bad. He fed his mind with bad thoughts at the expense of his professionalism, communication, and skills.

John was in the exact same situation as Frank. However, he had growth mindset, he wanted to learn, and he wanted to form meaningful connections.

It is not common that someone is in a position after 1 month to openly talk about expectations. John's manager was also more cooperative than usual, but he had very good reasons to cooperate.

People will support you, if it is worth for them to do so. Managers tend to guard budgets. They often have enough playground to keep the best people.

In some cases, salary ranges are fixed, and you can do nothing to earn more than the top of the range belonging to your position. For instance, a well-known bank has a policy in place that there is exactly 1 day per year to get a raise or a promotion. The key mindset there is to make sure you will be the one who gets promoted.

As a metaphor, imagine a good wolf and a bad wolf in you. Whichever becomes louder will shape your message. The key realization in managing your emotions is that you are the one feeding both of your wolves. Stop feeding the bad wolf. It almost always pays off to assume positive intent.

Eliminating Limiting Beliefs

You might need a therapist to conduct a session with you if the problems lie very deeply. In one period of my life, I had a therapist, who showed me the method, and we worked on eliminating one of my limiting beliefs together.

I was so happy with the result that afterward I wrote down 50 other limiting beliefs I had and kept on rereading the conclusions. Whenever a situation surfaced when my limiting belief came to my mind, I remembered how much I was convinced that this limiting belief had no foundations, and eventually, I instantly felt better. After a while, some of these limiting beliefs did not even pop up in my life.

It also helps to ask the powerful question of why. Everything happens for a reason. The cause why you feel miserable may also have a cause. Digging deeper and deeper lets you discover what you really need to work on.

I will show you a method in this section, based on the research of Albert Ellis. The name of the method is Rational Emotive Behavior Therapy.

His model, known as ABCDE, can be used to eliminate your irrational beliefs. I will illustrate the ABCDE model (Ellis and Dryden, 1987) through examples.

- **A** stands for **activating event** or **adversity**. This is the challenge you are facing, or a trigger activating your internal narration. For instance, you sent a performance review request, and your manager failed to reply.

- **B** stands for **belief**. You believe your manager is ignoring you on purpose, as you are just a replaceable employee. Before your feelings take over, try to record these beliefs.

- **C** stands for **consequent feelings**. In our example, you may feel neglected; your pride may get hurt. You may feel weak, and you may think your presence is not even important. Your manager must be judging you; after all, you are just a slave he can exploit.

- **D** stands for **disputations**. Are these beliefs good for me? Do I have proof that my beliefs are true at all? All I know is my manager didn't reply. I have no idea what he worked on. Maybe he didn't even read the email. Thinking that he did this to me on purpose is neither helpful nor constructive. If there is a chance that the above belief is false, the feelings I experienced are simply not justified.

- **E** stands for **effective new beliefs**. I believe my manager is competent and has good intentions in general. I also believe he is very busy; and even if he had time, we are all humans, and we have the right to make mistakes. Therefore, I will have a great, relaxing evening, and I will talk to my manager once he comes back from holiday.

Obviously, if you know that your manager has bad intentions, the belief that he has good intentions is a mere lie. Make sure the new beliefs suit you, but never ignore contradicting evidence.

Even though the ABCDE model is not perfect and it takes time to monitor your beliefs, it is a great tool to observe yourself, record your internal narration of events, and stay calm, cool, and collected.

There are situations, when the ABCDE method cannot be used. For instance, if your mindtalk starts taking control during a meeting, you won't have time to go through the whole process. The best bet you have is to breathe slowly after catching your mindtalk. You can reset your emotional state this way while sending your mind the signal that you deserve to take your time.

If your mental state is generally disturbed and biased, you also lose the ability to observe yourself objectively. Therefore, it is generally advised that a good therapist takes you through this method.

Exercises

1. Identify three limiting beliefs that are holding you back. Write them down. Use the ABCDE method to come up with effective new beliefs.

2. Record your negative mindtalk for a week. What triggers put you in the box during work?

3. Use the ABCDE model on triggers that put you in the box.

4. Reread your work on questions 1 and 3 daily for a week. Once you are done, the week after, copy only the effective new beliefs in a new journal and read them daily for a month.

5. Decide on what method you will use to interrupt your mindtalk on the spot. Practice this interrupt technique next time you realize your internal narration becomes destructive.

Face Your Fears

Goals of this section:

- Understand why narcissism is not the answer to insecurities.

- Formulate your opinion on perfectionism.

- Observe your advisors: the angel and the devil.

- Recognize and shut down defense mechanisms when fear takes over.

- Learn how to face your fears.

We left off with emphasizing the importance of eliminating limiting beliefs. If you pay attention to your everyday life, you will start believing that any belief can be challenged. The next logical step is action.

Action can be very hard at times. Has it ever happened to you that you know you should do something, you know what you should do, you know how you should do it, and you find yourself paralyzed? This experience is fear. Leaving your comfort zone is not comfortable.

This section is all about recognizing and facing fear. Remember when you are at the negotiation table, no scripts will help you if you find yourself paralyzed. Many people enter salary negotiations with a well-designed strategy, just to find out that the actual negotiation resides somewhere in their panic zone. The result is crashing and burning. In order to avoid this, you have to do your homework before negotiation starts.

Working on your fears is like debugging. Maintainability of your life is like the maintainability of your code. Similarly to software, your life may collapse if you settle for symptomatic treatment and avoid the root cause.

Neediness and Narcissism

"Listen, Mr. Smith. I really need that raise. I have a family to take care of, and, you know, rent has increased too, and I have not received a raise for 2 years."

This is one of the worst things you can say. Fear took over your thoughts. Fear is nothing else but an internal vacuum. Given that our hero does not think he has much to offer, he has no other choice but to appeal for empathy.

Beyond not showing any value, the worst thing about this approach is sounding needy. People rarely get anything from life, when they appear to need the result. People often get the biggest rewards, when they don't need them.

■ **Note**   Emotional fog spread by popular coaching and mentoring businesses often makes insecure people believe that neediness is a bad thing and should be eliminated. They invent a magic pill that you have to take to "combat neediness by working on yourself and becoming better." Unfortunately, this approach often lacks self-acceptance, and therefore, no real progress can happen. In my opinion, neediness is neither good nor bad. Neediness is just a mere symptom that our needs are not being met. The solution for becoming non-needy is not to cover our needs with compensatory behavior, but to meet our needs in ways we can influence. Getting a raise or a promotion does not address the insecurity of feeling that we are not enough without external recognition. By addressing the root cause of these insecurities, we can experience our self-worth and raise our self-esteem, and neediness will fade as a side effect of not having unmet needs anymore.

Once your mind tells you that you deserve what you are capable of negotiating, your actions will shift from being needy toward growing as a person and adding value.

Let's see the other extreme behavior:

> "I have closed 258 tickets in the last 1 year, which is 35% more than the second most productive person in the team. I took charge of writing meeting agenda, and this way, our meetings took 50% less time. I am the only one who knows how to handle this old piece of crap in our codebase, so I think it is time that we start talking about the evident gap between what I am earning now and what my expectations are."

Obviously, this example was just made up based on my experience of having talked to many people who felt entitled to get a raise and criticized and blamed their managers, their subordinates, and their peers for not recognizing them. Most people are a lot less direct in reality, but essentially, they tend to communicate that

- They are superior in some sense to their peers.

- They are the main reason why their team is performing better.

- They are irreplaceable.

- The company is taking advantage of employing them cheaply, because the company does not recognize their true genius.

Welcome to the world of **narcissistic confidence**. You appear narcissistic, when you focus on presenting your skills and achievement instead of focusing on doing good things in general. Narcissism promotes self-grandiosity and superiority and often comes with arrogance. Narcissistic confidence seeks for audience, and action in front of an audience is significantly different than action when no one is watching.

Real confidence comes from high self-esteem. High self-esteem encourages right actions, right actions deliver the right results, and results speak for themselves. This is why confident people appear more humble than narcissistic ones.

People recognize self-confidence only when they see it. Yet, most people have no clue about how to build it.

Raising self-confidence is based on raising self-esteem. Work on the six pillars of self-esteem in your life, and eventually you will become self-confident.

Narcissism is a compensation for the lack of self-esteem. Narcissistic behavior is driven by fear and will collapse under pressure. Positive reference experiences earned through shortcuts, manipulation, and luck tend to amplify narcissistic traits. This is when the perception of the individual becomes distorted, as the narcissistic image becomes distorted from objective reality.

Narcissism can be damaging in two ways: First, a narcissist may **compensate** for their own insecurities. Second, by getting used to the rewards of deceiving others, the narcissist may become a **predator** motivated or fueled by consuming narcissistic supply from others who give them attention, rewards, and encouragement.

According to my method acting coach, Shredy Jabarin, narcissists study good character traits to mimic them even though they have neither the emotional wealth nor the interest to possess them. It's like getting obsessed with the color red while you are color-blind. You surround yourself with objects that are red, but you don't see these objects; you just pretend that you appreciate the beauty of red objects. You carefully study the feedback of others who actually see these colors so that you can collect the attention and appreciation of others. This act can develop into a cluster B personality disorder that may or may not be curable.

In many areas of life, narcissism is a more successful approach than being needy. Yet, the approach is so fake that it won't be successful in the long run. As people spend more with a narcissistic person, they figure them out quickly. Most software developers work in teams, and narcissism works against cooperation, because the interest of a narcissist is not to cooperate, but to collect narcissistic supply.

▍**Note** Handling narcissistic people goes beyond the scope of this book. The author has had experience cooperating with multiple individuals in disarming narcissists both in a professional and in a personal setting. Cut all narcissistic supply, because narcissists are needy for recognition. The motivation to seek attention will eventually force them into mistakes. Any emotional exchange in front of an audience benefits the narcissist who has perfected to extract narcissistic supply from the audience watching the emotional exchange. Precise fact-based communication is essential.

Years ago, I believed that self-confidence is a calibration exercise between neediness and narcissism. Back then, I did not know the true meaning of narcissism. The path of appearing needy, then narcissistic, and then needy again due to not being able to take feedback is by far not optimal and may lead to a process of never becoming confident, but just appearing confident.

My current point of view on this subject is that you become non-needy if you learn how to meet your needs. Truly meeting your needs implies that you also get familiar with your shadow self. Your shadow self is the part of you that suffers when you get triggered. You may be selfish, jealous, bitter, frustrated, or resentful. To gain emotional immunity, study these emotions, and reflect on what the positive intent could be behind these emotions, because ultimately, all these negative emotions are there to protect your survival. Once you focus on this positive intent, you will find out that you don't need these negative emotions anymore. This is a long therapeutic process though, and **grieving** is essential to set yourself free from these limitations. This is why I recommended grief work and therapy instead of a solution I can sell you myself that would create an illusion of progress.

Perfectionism

Many people think that perfectionism is a buzzword that does well in resumes. I have seen many perfectionists stuck in their careers, failing to meet reasonable target dates and sometimes even failing to finish a learning product before its topic got outdated.

While I don't encourage people to be lazy, the dark side of perfectionism is that in reality, nothing is perfect, so you never feel that you are ready.

We should all stick to high standards in work, and in life. These high standards will never be perfect. Once you meet your standards, you have to calm yourself down saying, "I'm done; it's good enough."

The cause of perfectionism is fear – the fear of not being good enough, the fear of others laughing at our work, the fear of disappointing other people, or the fear of not showing our very best at work. In high school, I was a perfectionist. When I could solve 99.9% of all possible test problems, I worried about the rest. After all, what happens if the whole test will be about that 0.1% I might have missed during reading? This approach was very unhealthy, and not only my social life suffered from it but also my well-being.

If you are afraid of delivering bad work, develop quantifiable standards. Whenever you meet your standards, tell yourself that you've done a good job and it's good enough. Don't let the fear of perfectionism take over your life.

When you negotiate, don't talk about how much you value perfectionism. It would imply that you are afraid of getting things done.

Expectations of Others

Have you ever seen cartoons where the main character had a moral battle? There is always a little devil, telling the hero what to do. This devil is in constant battle with an angel. The cartoon character in between had to decide on which voice to listen to.

These moral battles happen on a constant basis.

The devil often represents our **instincts**. It seeks immediate gratification. You want all rewards here and now.

The angel represents your **super-ego**, responsible for moralizing. The super-ego manifests expectations of your family, your colleagues, and society in general.

The **ego** stands in between, trying to decide whether the devil or the angel should take over. The ego's job is to find balance.

Not meeting expectations of others results in pain, often as strong as physical pain. This is how the pressure of our society can condition us not to do anything extraordinary.

Your angel may tell you asking for a raise is not appropriate. After all, you are here to meet expectations of others. Your devil may tell you that you should either not work until you get a raise or you deserve a raise even without adding value. Needless to say, one force makes you needy by going against what you truly want, while the other makes you narcissistic by meeting your needs in a disempowering way.

You need to pay attention to your inner forces and free yourself from both your instincts and expectations of others. Your reward is a life free from distractions, and you can let go of fear.

Defense Mechanisms

When fear takes over, we tend to apply unconscious **defense mechanisms**. It is very important to realize that the drive behind these defense mechanisms is fear. Instead of combating the symptoms, the cause has to be fixed.

Procrastination is one of the most well-known defense mechanisms. Taking action is more painful than not taking action. No productivity management training will let you overcome procrastination. Taking responsibility and combating fear will do miracles.

Rationalization convinces you to find an excuse instead of taking action. Our super-ego tells us to avoid conflicts. Major problems in your company can be solved later; everyday routine is always more important.

Projection is about projecting your own doubts onto your colleagues and leads. These doubts can ruin your professional life. You may think the company wants to exploit you. Your peers want to climb the corporate ladder at your expense. People going out for lunch have to talk about you all the time. If you asked for a raise, your manager may even get you fired. All your colleagues are procrastinators; they just got lucky. When any of these feelings take over, you start concentrating on wrong things, instead of focusing on what really matters.

Blaming others. It is very easy to tell yourself others are less competent than you are. Whenever you point fingers, your index finger to be exact, remember that three fingers point back at yourself. Identify things you can influence, identify things you can't influence, and decide which group is worth focusing on.

Apathy convinces you that you have no enthusiasm to pursue something. You pretend through rationalization that your goal is not even that important. You want to get ahead in your career, but there are more important things in life, for instance, your favorite TV series, your favorite computer game, or just spending another half an hour on Facebook.

Fear Is a Symptom

The faster you run away from your fears, the faster your fears will run after you. Therefore, it makes a lot more sense to stop running, accepting where you are right now, knowing where you want to be, and taking action.

In some cases, your biggest fears should not be combated on the spot, as you would be sent to your panic zone right away. If you feel the challenge is far too high for you, record your thoughts, and develop smaller challenges that will eventually enable you to grow. Growth will increase your chances of facing anything you are afraid of.

When walking away from a fearful situation, make sure you never leave the stage without taking some action. This action can even be as little as going to a market nearby and starting negotiation on the price of a dozen of eggs. Make sure you also plan how you are going to face your fear in the long run.

Facing your fears is very much like a video game. If you have ever played any games in the past, you know that the first levels are often very easy. These levels are there, so that you can get a feel of what the game is like. For instance, in the video game *Sonic the Hedgehog*, all you need to do to pass the first level is to hold the right button and jump a couple of times. At first, even this action seems challenging. After a while, the first stage becomes routine for you, but there will always be a level on the game that appears to be too hard. Instead of giving up, you keep trying, and eventually you are capable of finishing the whole game with this mindset.

Taking care of your career is not different at all. The key to solving harder challenges is through tackling easier ones with managed stretches and avoiding the panic zone.

We have already seen that stretching can happen in two ways: through emotional growth and through action. Assuming emotional growth is a continuous process, your job is to manage your actions such that you spend as much time in the learning zone as possible.

Notice that most of the changes happen in a training environment. If you train your muscles, you don't start by trying to run away from a hungry lion. In a training environment, you apply **training mindset**. This is when you have time to fine-tune your performance and you have nothing to be afraid of.

Once you have to deliver, you are in **trusting mindset**, trusting all the skills you have acquired. When you are in trusting mindset, you are not worrying about whether you are good enough. When the lion comes, you are not going to focus on optimal breathing.

Exercises

1. Write down your fears related to your career advancement. Accept that they are there.

2. Have you ever felt needy in your professional career to get a specific result or acknowledgement? Find out the root cause for this behavior.

3. Define your standards. Observe what happens to you once you meet those standards without worrying about the expectations of others and without meeting your needs in a disempowering way due to temporary weakness.

4. Observe yourself for a week, and note down all sources of

   • Procrastination

   • Rationalization

   • Projection

   • Blaming others

   • Apathy

5. Find the root cause of your emotions developed in question 4.

Escape the Motivation Trap by Strengthening Your Willpower

Goals of this section:

- Learn why leveraging motivation at the negotiation table does not pay off.
- Find out the relationship between motivation and willpower.
- Learn how you can strengthen your willpower.

Think about your past and present colleagues for a moment. How many of them were equally happy on Mondays and on Fridays? One day, when my colleague in a key position told me that I should be happy because it's Friday and I could go out after work and get drunk, I already knew that something was very wrong with his mindset.

When the same person was about to leave our company, we figured out that he did nothing more than sharing animated gifs in our communication channels, browsing the Internet, and bragging about his past accomplishments.

We can conclude that many people tend to lose motivation once they are about to leave a company. This mindset is so human that it can happen to anyone, from customer care agents to managing directors.

One of the most extreme forms of trouble with motivation is cultural. What is May 1st about? In many countries, we celebrate Labor Day. We celebrate work by not working. In some countries, this celebration even has an undoing effect, namely, Luddism, that is, the destruction of equipment, vandalism, and some other forms of violence.

Some people keep expressing that they lost motivation, while others just get things done. Ask yourself which group you want to belong to.

Motivation at the Negotiation Table

Many people tend to leverage their willingness to motivate themselves to get a better offer. They say, "I will be motivated a lot more once I earn $10,000 more. As long as I am exploited, I will work less."

These people tell themselves employers and managers exploit people and you are a coward if you believe you should still give your best. This mindset may ruin your career. Do the exact opposite! Deliver your best, be a role model, and see things change around you. If you treat your career as a business, you soon figure out that you are responsible for getting ahead.

During your negotiation process, arguing about your motivation is equivalent to blackmailing your employer that your work is going to be less efficient if they don't give you more money.

You may earn an extra dime today, at the expense of future dollars. The aim of this book is to give you a development path to lasting solutions.

As you are reading this chapter, you may think that this sounds very nice on paper, but we are humans, not robots. Partying, having fun is important, and after all, show me people who are motivated all the time. You will soon see that you are partially right: no one is always motivated at the workplace. The art of getting ahead and enjoying the process is about getting things done, regardless of fluctuations in our motivation.

Motivation and Willpower

Many people mix motivation with willpower when they say, "I need motivation in order to complete this task."

Motivation is a reason for acting in a particular way. This reason is often emotional. You are motivated to do something, if you feel like doing it.

We use **willpower** from the aspect of self-control. It is the ability to control your emotions and behavior, in the face of temptations and impulses.

When saying "If only I had more motivation," people mean willpower. It is not always possible to motivate yourself. Motivation is all about feelings, and we have very limited control of our feelings through our emotional interpretation. Willpower is the ability to act even if you don't feel like it.

When thinking about success, willpower predicts success better than your IQ. You have a strong willpower if you show up and work effectively every single day, regardless of the circumstances.

Interrupting your work decreases your ability to act. Your thought process is often interrupted by smartphones and the Internet, and your ability to concentrate fades. Some tasks require you to deeply focus. Decreased concentration has a measurable effect on willpower.

As entertainment is so easily available, many people tend to check their emails, Facebook, WhatsApp, and their favorite personal news sites on a regular basis. If there is nothing else to check, there is always an animated gif or a meme around.

Willpower is just like batteries. Once your willpower is depleted, you are not able to take decisions anymore and will give up on taking action. This leads to procrastination. Procrastination leads to a decreased self-esteem. This is the process people are in, when they say that they have lost motivation. Talk to a person about to leave your company, and you see traces of this process.

People with low willpower can hardly offer much at the negotiation table.

Strong willpower does not mean that you have to be hard on yourself. Being hard on yourself on a regular basis is not only neurotic, but it won't allow you to enjoy the fruits of your effort. You get used to being hard on yourself, and you will always look for the next step without enjoying the process. Remember it is not the goal that makes the difference in your life, as enjoying the process is more important.

In fact, if you beat yourself up on a regular basis, sooner or later, your system will give up and you will rebel against too much discipline. You have to know the purpose of why you do things you do. This is why we will work on our goals and our purpose in the next chapter.

Strengthening Your Willpower

We have learned that willpower is a finite resource. Decision making depletes willpower. Eating, sleeping, hobbies, meditation, and the right holidays recover willpower.

Strengthening your willpower is also possible. Set your life up in a way that you drain your willpower the least.

Regarding your interruptions and notifications, I suggest turning them off during work unless you are accountable for answering within a specific time frame. Depending on your personal lifestyle, you might live a happier life turning them off privately too. Checking your private email every 5 minutes because of the fear of missing out on the latest piece of information is very unhealthy in the long run.

Once I was waiting for my plane. My good old phone was on my lap. As I checked something in my hand luggage, my phone suddenly fell down, landing on its touchscreen. The touchscreen got broken, as the concrete of Berlin Tegel was a lot harder than the Gorilla glass. I had to spend my 2 weeks' holiday without my phone. I had one of my best holidays ever. Instead of distractions, I focused on people. Since then, all my notifications are off, and people wonder why I don't pick up the phone, as my phone does not even vibrate when I get a call.

During work, define your maximum response time, and check your emails accordingly. Ask your colleagues to approach you directly in case of an emergency. Mute your chat notifications. Skype and Slack can both be used without notifying you every minute that something happened.

Forming **habits** is another way to keep your willpower at its maximum. Habits are decisions that you made in the past. When executing a habit, you don't have to decide anymore. At first, we shape our habits; then our habits shape us.

Getting stuck in a decision drains willpower. In an edge case having low impact, any decision is better than no decisions. Just decide, and move on. Should open braces start on the same line or on a new line? Our vote was 6 vs. 5. Frankly, I don't mind either way. I just invested time in counting the votes.

Entrepreneurship is all about validating assumptions. Many assumptions are proven false. You can aim better based on learnings of your previous action. You keep on moving; you keep on learning. You think of yourself as someone who puts things into action. Your willpower will not be depleted that easily.

When it comes to your self-image, put your identity at stake whenever you have to act. For instance, if you think of yourself as someone who never gives up a challenge, you keep on solving a hard task, until you succeed. If you think of yourself as someone who will create a lot of value in his career, you are not going to stare at your screen for hours, thinking that you are blocked. You will go and unblock yourself. This attitude extends the boundaries of your willpower.

When it comes to performance, willpower trumps motivation.

Exercises

1. What are the areas in your career where you would need more willpower?

2. What distractions do currently exist in your workplace? Which of these distractions can you eliminate starting tomorrow?

3. Check all the apps on your mobile phone. Which of them send you notifications? Are there apps that are worth uninstalling? For the apps you keep, consider turning off their notifications.

Summary

This chapter covers the most important aspect of your career: yourself. Most problems arise not because of not knowing what strategy to apply, but because there is something inside you that doesn't let you negotiate with integrity.

Integrity starts with happiness. We concluded that our goal is to work on our self-esteem. We identified some obstacles in your path.

Comparing yourself to others is dangerous, as it damages your integrity, and your self-esteem suffers. Compare your present self with your past self instead.

Improving your life by pursuing goals requires continuous effort, and this effort is often uncomfortable compared to the alternatives available to us.

We covered the differences between hedonic happiness, that is, the happiness of the senses, and eudaimonic happiness, the well-being of the mind. Ancient Greek philosophers and modern researchers in human psychology claim that eudaimonic happiness is the only lasting form of happiness.

After examining both forms of happiness, we identified that the key to successful negotiation is high self-esteem and a proper self-image. Your self-esteem grows whenever you take responsibility of your actions and your career.

We identified the six pillars of self-esteem based on Nathaniel Branden's work. These are consciousness, self-acceptance, self-responsibility, self-assertiveness, living purposefully, and personal integrity. If you work on these six areas of your life, your self-esteem will grow.

We compared fixed mindset and growth mindset. We collected the advantages of growth mindset and found that it is essential to put yourself in growth mindset, if you want your salary to grow accordingly. People with a fixed mindset will end up earning significantly less in the long run. Therefore, accept where you are right now, and treat your mistakes as learnings.

The number one obstacle of your growth is fear. Fear influences your actions by holding you back. You can only act with integrity, if you face your fears instead of avoiding them.

We covered two approaches that are far from being optimal. Needy employees are the least efficient negotiators. Employees with narcissistic confidence are a bit stronger, but their strength is based on their own fears. Although narcissistic people tend to be more successful than needy people, the reward of narcissistic effort is bitter without the developed senses, and the eventual self-destruction is often inevitable.

Perfectionism is another consequence of fear. Make sure that your high-quality work comes from sticking to your standards and not from the fear of not being perfect. No one is perfect, and perfectionism eventually leads to loss of motivation. Loss of motivation results in procrastination.

Fears hold you back in several ways. Defense mechanisms defend the subject of your fear from being worked on. Procrastination, rationalization, projection, blaming others, and apathy are all there to make sure you choose the path with the least resistance. By choosing the path with the least resistance, we keep our fears that are there to give some weight to our lightweight willpower.

Limiting beliefs can amplify our fears. We compared two mindsets: one in the box and one out of the box. In the box thoughts were identified as limiting beliefs. The ABCDE method of Albert Ellis allowed us to overwrite our limiting beliefs, leading to a more balanced life, and better negotiation skills.

We concluded this chapter with the difference between motivation and willpower. The most important takeaway from this section is that lack of motivation should never be used as an argument at the negotiation table. We concluded that people capable of managing their motivation by exercising their willpower often create more value to the company.

Please take the exercises at the end of each section seriously. The exercises at the end of each section tell you what you need to work on on a regular basis. As a result, you will need to rely on less tactics; and you will develop the necessary problem-solving skills and judgment to approach situations where simple solutions don't exist.

Discover Your Individual Goals

Focus Is Power

The goal of this chapter is to understand your current situation, understand where you want to be, and construct a plan to make your way to the desired destination.

To execute this process, you have to know what is available to you. A lot of people have no idea about what each career path holds for them. When I was inexperienced, I was one of these people. Had I known what the future held for me, I would have made better decisions.

We will start with exploring some possible career paths. This examination will be based on **career capital**. Your employer invests in your skills, competence, and connections to level up and create more value with your work. Everything that increases this value is your career capital. Career capital does not equal know-how; it goes beyond the skills you use at work on a daily basis. Career capital includes your professional network, your habits focusing on creating professional work, and even your personal brand.

If you could take away just one thing from this book, always focus on increasing your career capital. Negotiation techniques increase the rate of exploitation of your earning ability, but your base earning ability is directly proportional to your career capital. Negotiation techniques may create the illusion of progress,

© Zsolt Nagy 2019
Z. Nagy, *Soft Skills to Advance Your Developer Career*,
https://doi.org/10.1007/978-1-4842-5092-1_3

especially if you had made bad deals in the past. Real growth happens on career capital level though.

The book *Outliers,*[1] written by Malcolm Gladwell, emphasizes the 10,000-hour rule: "The idea that excellence at performing a complex task requires a critical minimum level of practice surfaces again and again in studies of expertise. In fact, researchers have settled on what they believe is the magic number for true expertise: ten thousand hours." Even though I personally disagree with allocating the constant of 10,000 hours to becoming excellent, I view the main takeaway from this quote is that it takes a lot of time to build up career capital.

Please also consider the quality of time spent according to the 10,000-hour rule. Random activities do not get you anywhere. The 10,000-hour rule is about 10,000 hours of *deliberate practice*. Just spending a lot of time on a subject does not make you better. Creating a deeply immersive experience opens up your channels for growth and learning.

In his book *So Good They Can't Ignore You,*[2] Carl Newport argued that you will become happy and passionate in your job as you gain valuable experience. According to studies he based his research on, job satisfaction has very little to do with passion you feel toward your job before you start working.

I personally find this advice slightly flawed. Most points of view on the extreme edge may lack the fluidity to map an individual reader's situation, and therefore, I tend to use extreme advice with care.

I strongly believe that hustling in itself is not a way to lasting happiness and fulfillment. While you can become passionate about what you do by getting to the cutting edge, you can also forge your passion before you spend your 10,000 hours of deliberate practice on the subject. The factor that truly matters is execution.

Passion is obviously not enough. I have a great friend who is obsessed with racing simulator hardware. He builds better equipment than the companies selling inferior products for more than a thousand dollars. However, apart from one arcade game hall, no one knows about his developed talent, because he is not marketing himself. He uses his cutting-edge equipment to finish in the midfield of online simulators, because he treats this as a hobby.

Increasing career capital and deliberate practice is not enough to excite passion. How many people are there who live an unhappy life believing in the myth that you have to grind and hustle for years?

---

[1] Malcolm Gladwell, *Outliers: The Story of Success* (Little, Brown and Company, 2008).
[2] Cal Newport, *So Good They Can't Ignore You: Why Skills Trump Passion in the Quest for Work You Love* (Grand Central Publishing, 2012).

Oftentimes, the result is an unfulfilling life. Being hard on yourself for years to achieve does not guarantee that you will benefit from the goals you accomplish. In exchange for pursuing a goal you don't want, you get to miss out on life experiences and career opportunities that you truly want your life to be about. Is this a fair trade?

Think about education, for instance. Coding bootcamps have become popular lately. These bootcamps attract many students who studied another profession, oftentimes abandoning a career path their parents selected for them.

In his book *The Third Door*,[3] Alex Banayan gives you his own example of how he confronted his fears of telling his parents that he would opt out from his college education to pursue his own dreams. After reading the book, you decide if this pursuit was successful. His goals changed throughout the journey, but the experiences he gathered along the way exceeded all my expectations as I read the book.

Alex followed his dream and executed his pursuit beyond imagination. He is definitely not a target of Cal Newport's passion hypothesis. Those who dream but fail to execute properly are the real targets.

Let me contrast this path of a mentor; let's call him Tom. Tom had a tough childhood, and at the age of 20, he had to finance his university studies and his family from his own pockets. He launched a consulting business. He appeared really professional, engaging, and likeable in his marketing materials and grew his social media presence fast. Many of his followers resonated with the ideas he was preaching. One idea was that in order to make real progress in life, you have to have a "pain period" to pay the price of success.

I followed his progress around the end of the second year of his 5-year pain period masterplan. I even booked individual and group coaching with him, because I needed professional help in the area of his expertise. During the individual sessions, it was evident to me that he was really hard on himself; but he behaved professionally, and he just communicated that his expectations were very high toward everyone he worked with.

I attended a live 3 days' workshop on a topic that I thought would help me. I greatly benefited from the structure, but I noticed some people did not share my point of view. They were frustrated, because they felt they were treated like trash. Their questions were answered with passive aggression and hubris, and some parts of the workshop were not even delivered.

---

[3] Alex Banayan, *The Third Door: The Wild Quest to Uncover How the World's Most Successful People Launched Their Careers* (Random House, 2018).

As Tom did everything on his own, near the end of the workshop, he was mentally exhausted and started criticizing some of the less disciplined members. Then he changed the curriculum of the workshop saying that he already overdelivered in other areas. I was happy with my own experience and progress, so I was puzzled as to why the others were so unhappy.

Fast-forward 3 years later, I was curious what he was up to and checked in on him. I learned that he had earned more than a million dollars, and on paper, his clients gave him an almost five-star rating on his service as "verified buyers."

As I still had access to some of his communities, I observed the webinar launch of one of his products. His marketing was great. At the same time, in his webinar, he appeared arrogant, acting as if his audience should be grateful that he was willing to share his knowledge with them. His perception got detached from objective reality so much that he did not even realize that he lost a lot of people with his communication. As he was advertising like crazy, he still made a lot of sales.

I ran into Tom by coincidence one day and had a chance to catch up with him. As we talked, I noticed that he gained some weight and he appeared tired and unhappy. He was still proud of what he achieved, but said it was time to end the pain period, because he now had the skills to market himself for the rest of his life. At the same time, he didn't make an effort to empathize with anyone around him. His perception got distorted, and he could not connect with objective reality.

There is a major cost associated with his life choice of hustling and grinding for 5 years to fuel his need for significance and certainty. His path resulted in an altered perception of reality, which taxes not only his professional future but also his private life. This is the reason why I included Chapter 3 on emotional balance so that you build your career on solid foundations and you get more clarity on what you want from life.

On my web site, I've created a course where you can receive personalized help from me tailored to your own situation to discover your purpose and set your life up for success on your terms. For more information, check out `http://devcareermastery.com/career-breakthrough-program/`.

The ambitious goal behind this program is that you receive personalized help and coaching based on what you want your software developer career to be about.

Maybe you are employed, and you want to start your own business. Guess what, there are people out there who have done it with less coding skills than you do, launching a simple WordPress site with no backup, no firewall, nothing. And they made a good six-figure income. Members of the Software Developer's Career Breakthrough Program have all this knowledge and more, so they make progress faster, because they are not constrained by technical difficulties.

Others want to be self-employed or want to work remotely, but they are afraid of leaving their job when they join the program. All we need to accomplish is to identify your ideal role, see what's out there in the market, identify the expectations, and find a plan to increase your career capital and design your personal brand such that you can present an irresistible offer and add real value to your clients or employers. Sometimes the solution is a lot closer than you would think.

Oftentimes, developers don't know what is wrong. They are just frustrated, because they want a better life, better working conditions, less technical debt, and less corporate politics; but they don't know what they want. This is why I included a process where your unconscious mind will guide you to find the answer.

If you have read *So Good They Can't Ignore You*, I suggest reading the book *Mastery*[4] by Robert Greene. In this book, you will find a lot of examples of how people found their passion in life. The phenomenon of becoming a master in a career that was not fulfilling emerged over and over in history. Greatest success was forged when real passion coming from deep within us met with mastery.

As a consequence, I am not advocating the follow your passion mantra, but I don't buy Cal Newport's argument about the demonization of following your passion either. I believe that everyone is different, and my job is to show you each side of the coin, instead of trying to prove a point about following your passion.

In the first section, we will look at different career paths. Then you will get some questions so that you can discover where you are right now and where you want to be.

Career Paths for Software Developers

Goals of this section:

- Discover how generalists become specialists and T-shaped professionals.
- Compare the tech expert, the domain expert, the process expert, and expertise in leadership and management.
- Intrapreneurship in startups and corporations.
- Consider career capital in shaping your career.

---

[4] Robert Greene, *Mastery* (Viking, 2012).

The abundance of roles in IT makes it hard for us to plan our careers. Whenever you find a title, do your research, and make sure you categorize what is at stake. I also suggest determining the career path your potential role leads toward.

In this section, we will examine four different career paths: the **tech expert**, the **domain expert**, the **process expert**, and the **leader/manager**. The enumeration is not exhaustive. You may find a different career path that works for you better. You will receive some guidance on which career path to target.

Independently from these career paths, we will also explore valuable attitudes. You will find out more about the **intrapreneur** attitude, which is somewhere between an employee and an entrepreneur in terms of ownership and taking responsibility.

If you were waiting for the definition of different architect positions or the difference between the manager and director levels or the difference between a senior tester and QA engineer, I have to disappoint you. There are so many positions out there that an exhaustive enumeration would not fit in this book, while a non-exhaustive enumeration would not be too useful. I encourage you to do your research. If you do not have an idea about what the responsibilities of a position entail, chances are that position is not for you yet. In other words, to become an architect, you have to work with architects first to gain enough experience.

Tech Experts

If you have gained career capital in writing code, software design, and researching solutions for complex problems, this is a possible career path for you. An expert has the knowledge to dive deeper into a topic than his or her peers. An expert specializes, while others stay generalists.

A **generalist** can solve many simple problems, without diving deeply into a certain domain. Imagine a generalist as a horizontal profile of knowing a lot of things on a basic level.

A **specialist** lacks the knowledge of a generalist, but has unmatched knowledge in their area of specialization. In other words, a specialist may be able to write client-side rich web applications using JavaScript, React, and Redux. They know the ins and outs of the frameworks and can write efficient solutions. However, when it comes to server administration, writing efficient CSS, or even setting up a developer environment, a specialist may crash and burn without the help of a generalist.

The profile of a specialist is vertical, symbolizing that they can drill deeply into one area, but have lack of expertise in a lot of generic things that are required for launching a product. I remember once an Android developer asked me about very basic HTML questions that all junior JavaScript developers knew. In the Android world, this person just hadn't needed HTML knowledge.

There is high demand for **T-shaped professionals**, having both a horizontal generalist profile and a vertical specialist profile in one field. T-shaped professionals can work on their level of expertise without needing many people to support them. Being a T-shaped professional is an act of taking full responsibility for your actions, and this is why I suggest that you become a T-shaped professional too if you are targeting an expert role.

The learning plan of an expert is shaped naturally. If you figure out that there is a skill in high demand, go for it, and learn everything you possibly can to become a reliable expert. Practicing your skills on the job and working on meaningful side-projects is a reasonable strategy.

In some occasions, good experts earn more than their leads. An excellent expert almost always earns more than a demotivated manager who wanted to become an expert but was tricked into believing that becoming a manager meant more money.

The main advantage of staying an expert is that you have already gained some career capital in junior- and mid-level software development work. Managers don't always stay involved in software development and have low career capital to offer in terms of leadership and management skills after switching from an expert path.

Experts may be forced into undertaking responsibilities that make them code less. For instance, as you deal more and more with architectural design and detailed design, you may figure out that you are not a coder anymore. I encourage you to still be a part of the development team instead of giving them high-level advice from an ivory tower. This is how you grow and contribute in a meaningful way.

Domain Experts

A domain expert is genuinely interested in how their industry works. The added value of the domain expert lies in knowing the ins and outs of the product or service a company is offering, including ideas about how to make things better.

Even with average coding skills, the observations of a domain expert may save a lot of work. Domain experts tend to shine when reviewing the specification and planning the implementation.

Eventually, domain experts become product managers.

A product management role is often slightly better rewarded than a software engineer role with the same experience. Product managers run the show; they create roadmaps, engineer requirements, and have in-depth knowledge about the industry. They also know how to interface with software developers.

If you find yourself excited about reading the roadmap of the product you are working for and you envision how users would use your product, chances are your future is in product management. Just make sure that you are not only passionate but also collect some career capital before making the transition.

If you read a specification and you immediately start thinking about implementation challenges, chances are you would be excited about technology too much to become even an average product manager. You would miss writing code, and you would miss solving hard problems. Sacrificing your career capital will make you not only unhappy but also less competent.

Process Experts

Many people mix product management with project management. Product managers deal with products; and project managers deal with time, cost, quality, processes, organization, and systems. Projects have a given start and end and have a purpose of shaping a product.

Regardless of the differences, there is still a tendency of software developers ending up in the domain of managing processes. They become scrum masters, design thinking coaches, or generic project managers.

Some of these roles require complex, generalist knowledge and a lot of emotional intelligence:

- You need people skills just like leads.

- You need skills on knowing the problem domains, just like domain experts.

- You need to be able to interact with tech experts, requiring you to have a generic understanding of even the vertical part of the knowledge of a T-shaped professional.

- You may need business skills to understand the benefit of stakeholders.

- In addition, you may need to know techniques of your own field better than anyone else.

In practice, process-level experts hardly meet these criteria. They are good in exactly one area, and they learn the rest on the job. As a consequence, instead of becoming a catalyst for productivity, their added value often decreases.

An excellent software project manager is very rare, as the role requires a lot of knowledge in different areas.

Leadership and Management

A leader is often a special T-shaped professional, where the horizontal part of the T is very thick and the vertical specialization is in emotional intelligence, communication, and people skills, sometimes coupled with project management and/or software design. In addition, if you lead a team of specialists, you have to have the same specialization as well. Your specialization will not be as deep as that of your best experts, but on some level, you have to be an integral part of your team. Instead of retreating to your ivory tower and watching other people solve problems, your job is to cooperate with other people.

A leader leads by example and gains the commitment of other people by motivating and inspiring them. A leader is there to serve the needs and wants of other people.

In order to be a good team lead, you have to be able to put yourself in other peoples' shoes and represent their interest if needed. The sphere of responsibility of a lead includes the whole team, and the immediate surroundings of the team.

Management is more about operational duties. Managers tend to create plans, set goals, organize resources, plan budget, and monitor their KPIs (Key Performance Indicators).

A manager executes processes and makes sure the people they are responsible for are functional. Leadership is more about people skills, while management is more technical and process oriented. In order to be a good manager, you have to be structured, persistent, and on top of things.

The bad news is many companies employ incompetent leads and managers. A lot of experts, product-savvy people, and generalists think that the key to career advancement and higher salaries is to become a lead. Nothing is further from the truth.

In the last century, your success depended on how many people you managed. This is an outdated concept. In the 21st century, experts often earn more than their leads.

The concepts of **management** and **leadership** are often mixed up. They are two different activities that complement each other. Regardless of whether you are called a manager or a leader, both sets of activities are required for success.

This is why the complaint of having too many managers but no leaders tends to surface on a regular basis. People with bad people skills or bad communication are just not meant to be leaders, for the sake of the whole team.

Many developers ask how they can collect career capital on leadership if they are not yet leaders themselves. Let me give you a couple of tips.

First of all, watch the TED talk *How to Start a Movement by Derek Sivers*.[5] It is often the first follower who makes a really good leader. Every leader needs reliable followers. Just watch what your leader does, and make your leader better.

Second, you can take the lead also as a follower, by the time you have gained enough career capital. Some people cannot even be responsible for themselves. Increase your ownership of your tasks, go beyond the horizon of what you do, and start coordinating your tasks with others. Highlight common-sense observations that your lead might have missed.

Eventually, your leads will back you up, and you will end up in a position where you can become a leader.

Intrapreneurship

Intrapreneurship is the act of behaving like an entrepreneur while working in a company. A company wins a lot by letting intrapreneurs practice their abilities, even though risks stay on the company's side. Intrapreneurship is not born out of the desire of getting a higher salary. It is usually a preferred, natural behavior of a few individuals.

When some people tend to say the magic statement "I refuse to do it; this is not my job," it is a good sign that you are not dealing with an intrapreneur. A true intrapreneur may still reject these tasks if they know that someone else can do the same job faster and cheaper. However, even then, a true intrapreneur would find that person and negotiate the details.

If you are not an intrapreneur yet and you would like to become one, the first five chapters of this book will be very beneficial for you. The key elements of becoming an intrapreneur are the following:

- Curiosity should be one of your top values. How can I solve a problem I have never encountered? How can I make this work? What can I do to measure what my clients see, feel, and think? These are typical questions of an intrapreneur.

- Apply common sense to everything. Common sense overrides processes, software development best practices, principles, and basically all rules that are there for a generic guideline. If your web application is broken, the rule of not deploying on Fridays may be challenged if the problem is very serious.

---

[5]www.ted.com/talks/derek_sivers_how_to_start_a_movement

- Have an eagle eye. Point out flaws, where other people don't notice things. Be ready to reject a specification if it contradicts to your own common sense. Know when another team is discussing a problem, and prepare to jump in if it's an important topic.

- Know what your leads and the stakeholders of your project have trouble with, and be ready to help them even if it is not directly related to your job description.

- Innovate, experiment, and lead by example.

- Have a good feel for costs and profitability. Developing a dropdown list for 2 weeks is a lot more expensive than buying the best available component. In all companies, there are activities that waste a lot of money for literally no return. Make sure you work on tasks that matter.

Imagine that on a Friday evening, the CEO of your company finds out that you are the only frontend developer around and he cannot reach the lead of the backend team. He tells you in a worried state that he just found a major security leak. He asks you to verify this leak and, if it is indeed there, shut down the whole service immediately.

Your job is here twofold. First of all, the best outcome is to prove that the security leak is not there. It is your decision how much time you spend on trying to prove this, but you should develop a feel for how harmful the security hole is within 5 minutes or so. If you don't have the time to research what's wrong, shut down the service, trusting the decision of the CEO.

Don't ask for confirmation or clarification of details. If you can give a reliable and quick answer to the CEO on what was exactly wrong and do the necessary firefighting yourself, then inform both the CEO and the lead of the server-side team on the circumstances and actions taken; it means that you have taken full responsibility for this task, even if it was not in your job description. As an added bonus, you may find a solution without shutting down the service. The answer "I don't know how to do it; this is not my job" is very negative.

Think about a similar scenario. What would happen if you discovered this error yourself and no decision makers were around? The answer is not obvious, and there is a very thin line between doing the right thing and doing something very wrong. Your task as an intrapreneur is to take full responsibility regardless and take educated, reliable decisions.

Similarly, intrapreneurs often have the ability to optimize company processes. Suppose there is repetitive manual work in the company, equaling 160 hours of work per month. An intrapreneur may research developing a software solution that automatizes 80% of the work which leads to a saved capacity of 128 hours per month, and the solution also provides additional scalability, without hiring anyone.

Intrapreneurs tend to become true entrepreneurs eventually. Most of the time, their employers know their ambition, but their skills are so rare in the 21st century that companies feel privileged to have them on board.

If you want to make a real difference in your earning ability, become an intrapreneur, show that you are committed to taking responsibility, and watch the positive consequences of being treated uniquely in your company. It does not matter if you are an intrapreneur expert, lead, product manager, or project manager; applying common sense on top of your duties is an extremely valuable skill to develop.

Generalists

Being a generalist is not a bad thing. You won't earn as much money as people taking more responsibility than you, but life is not only about work.

I have seen a lot of people who were completely satisfied with a comfortable job, even without salary raises. You may choose to live your life just after the 8-hour workday is over and take things easy. Contribution of generalists matters, and they should be a respected member of the team.

A generalist living with integrity and choosing a work-life balance biased toward family life, playboy life, or adventure is fully acceptable. Living with full integrity with your values makes you happier than pursuing more money or fame half-heartedly.

Even if you don't go the extra mile, some principles of this book still apply for you. Generalists also have career capital to rely on, and this capital may be more valuable to your employers than the amount of money they are paying you.

Build Career Capital and Discover Opportunities

Instead of following the illusion of earning more money, follow your instincts about what kind of work suits you the best. There are so many bad leaders, mediocre experts, and unmotivated project and product managers around anyway. Don't be one of them!

Your instincts develop as you gain career capital. The book *So Good They Can't Ignore You* formulates its fourth rule around this idea. The rule states, "Think small; act big." Always focus on mastering your current field of focus. As soon as you get to cutting-edge level, your horizon will widen.

The reason why there are so many incompetent people in key positions is well summarized by the **Peter principle**. When getting a promotion, the capabilities and earning ability of a person are perceived according to their performance in their current role. This is how people are promoted to a level, where they are not competent anymore.

So Good They Can't Ignore You even dedicates a rule on when to turn down promotions. Some people are promoted before they gain enough career capital. They get more control, but cannot live up to their responsibilities. This is very dangerous and should be avoided.

As you gain more career capital in your job, your long-term earning ability will be higher than in the case of accepting a promotion you are not competent for. Therefore, choose wisely!

Exercises

1. Are you a generalist, a specialist, or a T-shaped professional right now?

2. Are you currently on the path of the generalist, the tech expert, the process expert, the domain expert, or the leader?

3. What three things can you change starting today that will move your attitude toward becoming an intrapreneur?

Where Are You Now?

Goals of this section:

- Discover your relationship with your job.

- Be able to determine your status at the company you are working for.

- Evaluate your earning ability objectively.

- Take your current financial situation into consideration.

The history, background, and current situation of every single person is unique. In order to get to where you want to be, you need to know where you are right now and take action in the direction of your destination.

In order to improve, you have to accept your current situation and take responsibility for it. For instance, if you start sweating whenever someone asks you how much you would like to earn, taking responsibility is the only approach that helps. Denial or positive thinking won't give you any edge during your next negotiation. Starting small and improving your skills on a regular basis will make a difference.

The outcome of this section is a proper summary of all the circumstances that have an influence in your future. I suggest preparing pen and paper and thinking about your current situation. Colorful pens help you, as you might want to illustrate your answer with different colors. This is how you help yourself remember and think about your own situation.

We will later design where you want to be, and we will also work on your personal brand later in this book.

You and Your Job

The first group of questions are on your current and past jobs. If you are a student with no experience, write down the advantages your school may give you, and write down all projects that could be used for marketing yourself. If you already have an employment history, ask yourself how satisfied you are with your current working environment and your current role. This determines how urgent it is for you to change your situation. Know whether

- You love your job and you find earning more money a nice-to-have.

- You would love your job if you earned more, but the thought of earning a low salary has an effect on your daily well-being.

- You would love your job if you worked with the right people, but you can't stand some of your colleagues.

- You love software development, but you actually don't like your job, due to lack of challenge, too much stress, bad processes requiring you to work overtime, bad working environment, and so on.

- You feel the urge of getting a promotion.

- You are only a software developer because of money, and you just pretend liking your job, but you actually hate it.

If you have any problems, determine where they come from. You may have a problem with your employer, your current position, your current team, your current manager, or yourself. It is the hardest to accept if you cause yourself problems, as most people prefer blaming others. This book is for you, and you get the best results only if you are honest with yourself.

In some cases, there is no other way, but to leave your current employer. For instance, if you are not comfortable with the product or service your company is offering, your personal life may suffer. If the work you do is in conflict with your personal values, it may be time for a change, so that you can stand up for the cause you support.

It is also possible that the skills and connections you gain are so valuable that it is worth staying even if you don't earn a dime more for a year. The opposite may also be true. I have seen so many people claiming they are senior developers with 5 years of experience, just to find out that they relived the same I year experience five times. Your career is in your hands, do something meaningful, and make sure you take your earning ability into consideration.

As I am writing these lines, I can recall an interview for a senior position. The candidate looked all right, he introduced himself, and looked knowledgeable. I gave him a coding task. As soon as he started summarizing the task, I immediately knew that something was wrong. He just didn't have the necessary coding experience to function as a software developer.

Note Companies worth working for employ interviewers who can see through your attempts of trying to impress them. You cannot hide your seniority with tactics, rhetoric, Machiavellianism, narcissism, or other means. You can choose to work on your skills though to increase your career capital to the level where your services become attractive and marketable. If you are interested in finding out how you would do in a short tech interview, grab one of the limited available spaces for a free interview consultation.[6]

Your Status at Your Current Company

Figure out how you are perceived in your current company. Are you a superstar? Do people listen to your advice? How replaceable are you? Are projects heavily depending on your input? Are you just someone who comes in late, leaves early, and does nothing productive? Would your current team miss you? How well are you connected with your peers socially? Are others trying to climb the corporate ladder at your expense? Are they just helping you succeed? Does the CEO know who you are and what you do? When there is an emergency situation, do people count on your input, or do they rather avoid you?

Your negotiating power heavily depends on your perceived value, let it be based on social connections or performance or both.

In some companies, especially in banks, software developers tend to have low status. Businesspeople tend to have high status.

Once I worked for a company offering stock market education. It was run by a businessman, whose values were connected to entrepreneurship. Developers typically had low status, and they were just viewed as a necessary cost.

---

[6]http://devcareermastery.com/free-interview/

A company founded by software engineers creates the exact opposite experience. Developers have the highest status of the company. All developers are treated with maximum respect. Budgets are shaped such that the core operations of the tech company, that is, innovation and software development, prosper.

Even if your status is low, you can improve it. Your trump card in getting higher status is developing interest in what the business is doing. For instance, back when I was supposed to be a cog with low status in the machinery of the education company I worked for, I found out something in common with the business owner and myself. He had a huge respect toward people who can make money on their own. Given that I was quite successful playing online poker back in the days, I became the poker mentor of the founder of the company. My lead in the same company became a stock- and commodity trader. These commonalities helped us increase our influence. We did not do these activities because we wanted to impress the business owner. We did these activities, because we wanted to. As our values aligned with the values of the owner, we could cooperate better.

Your Current Earning Ability

Figure out how much you could earn right now with your current skill set. Visit job search sites such as glassdoor.com or payscale.com. Make sure you read job descriptions, and figure out how much you would earn if you applied for a different job right now. For instance, if you are earning $70,000 per year and you can see that you could earn $120,000 right now, you are in a different situation than in the case of earning your fair market value.

Most negotiation books try to give you uniform solutions regardless of your situation. This is why we take your own situation as the starting point: you need to take the exact steps that apply to you. Some of you will be able to skip steps of the negotiation process.

Note that money is not the ultimate answer when comparing positions. For instance, I could have worked for a company where even the hiring manager said I would have a lot of work. I would have had to travel more than 2 hours a day to reach their office and go home. They would have made me work at least 50 if not 60 hours per week based on my contract, for the prospect of maybe earning a bit more before I got my first heart attack.

I argued in favor of gaining career capital. Some of my readers may argue that continuous overtime is a way to gain career capital faster. Nothing is further from the truth. Quoting *So Good They Can't Ignore You*, "if you just show up and work hard, you'll soon hit a performance plateau beyond which you fail to get any better."

It takes post-traumatic growth to get better. It's like working out. When you lift heavy weights, great. As your muscles get repaired overnight, they grow after the big trauma, as your body figures out that you need more muscle. Companies that make you work hard for 60 hours a week are equivalent to a gym session where you keep on lifting light weights for a very long period of time. Your muscles won't grow.

Some companies are interested in providing a healthy, encouraging environment to develop your skills, to unlock your potential. The gain in career capital with these companies is very valuable.

If a company is trying to exploit you massively, be aware that a company with better **culture** exists out there; and as time goes by, more and more companies will compete for talent, as the gap between available positions and skilled workforce continuously increases.

Take your situation seriously, as you are responsible for your own future. If you hate your current job, as a software developer, you will always have a lot of options. Many employers realize that it is their interest to treat their employees well. For the rest of you working for tyrant consulting companies, you have some work to do.

Quantify every factor, and don't just accept the position paying $5,000 more. Examine, compare and contrast many factors, such as the following:

- **Number of days of leave**: I got 32 days of leave and around 8 days of public holidays in Germany, plus as many sick leave days as I need. I know, this is not normal for most countries.

- **Culture**: Does leadership absorb pressure, or they delegate it to you? Is there a lot of unhealthy pressure anyway? Is the company culture open? Does your opinion matter? These are all important questions. If a tyrant has exploited your good will for a year and often made you work 70 hours a week, you know that your health and well-being is worth more than the extra money.

- **Extra health insurance, gym membership, food, drinks, company car, company laptop, free apartment, and so on**: These often add small benefits. Make sure you consider their materialistic value, instead of being emotionally attached to your car, for instance. After all, you could buy a brand-new car with the money you get after a proper raise.

- **Prospects of promotion**: It is evident that in some companies, you will march forward, and your earning ability is massively increased as you gain more domain-specific experience. If you know that your salary will be increased by 50% in a year, it makes little sense to accept an offer for 30% more. Oftentimes, you don't even know if the other company will be around in a year.

- **Connections and networking**: Some companies employ potential company founders and people you can cooperate with in the future. Others employ *World of Warcraft* fans, and shield you from business. Never underestimate the value of your connections. One day, you might become an entrepreneur yourself. Your first client may be your current company.

- **Stock options and extra bonuses**: Most people overlook these perks, just because they are hard to monetize. Research what these perks mean to you in your context.

Your Personal and Financial Situation

Look at your current financial situation and lifestyle. Do you desperately need money because of loans, debt, a health issue in your family, or any other form of misfortune? If you need to repay a loan of $200,000 in 3 years, for instance, and you are earning $60,000 a year, you may need a short-term solution. Alternatively, you might just have a couple of dreams to finance your lifestyle. Postponing some of these dreams in favor of a long-term career goal is often justified.

There are simply phases in life when earning more money is inevitable. There are also phases when the best decision is to exercise willpower and pursue your long-term goals instead of a toy you want, but don't need.

I am not in a position to give you personal advice on life choices. Therefore, I just shift your attention by telling you that this matter is important. You may be in a situation when earning more money quickly matters. Alternatively, you may tolerate short-term sacrifices in exchange for lucrative opportunities in the long run.

Exercises

1. List all the advantages and disadvantages that you can find about you working for your current employer.

2. Determine your status, and determine the status of software developers at your company.

3. Determine your earning ability in the company and outside the company.

4. Consider your financial situation, and prioritize your next steps accordingly, let it be a raise in your current company or a new position.

Where Do You Want to Be?

Goals of this section:

- Design your career in a SMART way.

- Develop a 5–10-year perspective, and break down your goals backward.

- Put your plan into action right now.

- Be flexible in your approach.

Following the previous section, you should determine whether you are in an emergency situation or not.

In case of an emergency situation, desperate times call for desperate measures. For instance, if your current employer is exploiting you, you are earning below market value, you have debt to repay, and you have low status with no prospects of promotion, it is evident that you have to look for a new job right now.

If you are earning slightly above market value, you love the technology, you have high status, and your job provides you with inspiration and challenges, think twice about leaving. Imagine the only problem is that your peer is earning $10,000 more than you and you are jealous of his vacation in Brazil. Pursuing an immediate change in exchange for slightly more money might not be worth it. There are ways to overcome these bad emotions and transform your feelings into an internal drive that increases your career capital beyond your current imagination.

Handling emergency situations should always be done professionally. I still suggest reading the whole book. However, you have no time to make noticeable changes within days. There is no time to build your online presence or establish a long-term feedback loop in your company. Yet, you should be aware of all these factors and start working on them consciously, as soon as your emergency situation is fixed.

Even if you are in a desperate situation, it makes sense to plan your future and choose the best possible option. For instance, if you got fired with a notice of 1 month, it is still worth choosing the best possible option instead of accepting the first offer that comes.

Design Your Career

A significant part of your life is your career. Wasting your career is not a lucrative idea.

According to Tony Robbins,[7] a lot of people are caught up in making a living instead of designing a life. When they come to the end of their life, they realize they lived one-tenth of it. Not because they weren't intelligent, but because they missed the ability to take action and produce results.

Most people have no idea about what they want in life and only get what other people are planning for them. This may be a comfortable, low-paid job, where you have to do boring, repetitive tasks.

The most common form of goal setting people make is New Year's resolutions. This is a very ineffective way of goal setting, as most people don't take it seriously enough for several reasons. First, for most people, these goals are rather dreams.

Most professionals have heard about the **SMART** acronym. Although the meaning of each letter varies in the literature, it is worth concluding that goals have to be

- **Specific**: Focus on areas in your life that matter, and clearly define your goals.

- **Measurable**: You have to know when you make progress.

- **Assignable**: This part is obvious. As you are responsible for your career, make sure you target goals that depend on you.

- **Realistic**: If you are a junior developer, chances are you won't become a CTO in 1 year.

- **Time-based**: Set a reasonable target date for yourself.

Make sure you break down your goals properly. If they are not broken down, chances are the goal will stay too big for you.

Make sure you have meaningful goals that really motivate you. Most people never think years ahead. This is where New Year's resolutions fall apart. The goals are either not meaningful enough in your life to pursue, or they are too big to achieve.

---

[7] www.tonyrobbins.com

We always overestimate the change that will occur in the next two years and underestimate the change that will occur in the next ten.

—Bill Gates[8]

Even though having a 10-year plan is a bit too much, especially if you are just starting in your career, taking the next 3–5 years as a foundation will do miracles in your future.

For instance, when I graduated, I knew I wanted to work in an English-speaking country. I could have targeted the United Kingdom, but it was a lot more motivating for me to shoot for some more exotic countries, such as Australia, Canada, Malta, or even the United States. Given that I did not want to sacrifice my family connections, I ended up choosing Malta, which was just a 2-hour flight away from my home.

When I first read about Malta, I saw that salaries were quite low. On the other hand, by the time I got there, I already had more than enough career capital to make a difference.

My goal was not only to go to an English-speaking country. I also wanted to experience the Mediterranean lifestyle. As a university graduate, I was very much motivated by living next to the sea. I still remember how good it was to take a walk every day next to the sea and to work out in a gym overlooking the sea. At the same time, I was making progress in my career.

Even though I first investigated the chance of moving to Malta 3 years before the actual move, I only made the decision 1 year before I boarded the plane. Back then, I was already employed by a Maltese company part-time, working remotely from my hometown.

The summer before I moved to Malta, I spent 2 weeks there. Having won a poker tournament ticket and a room in an excellent five-star hotel and having secured an office to work in for the second week of my stay, I had more fun than ever before. At that point in time, my decision was obvious.

I spent 4 years in Malta. After the end of my first year, I already knew that it would not be the final destination of my life. I visited two other countries every year and gathered career capital in the specialization of frontend development and leadership. By the time I got to Germany, it was only a matter of half a year until I became a team lead.

---

[8] This quote appears in many different forms. This particular version is frequently attributed to Bill Gates in the afterword of the 1996 edition of *The Road Ahead*.

In the last chapter, I emphasized that having fun in your way is essential in your success, let it be partying, establishing your lifestyle, or pursuing your hobbies. In fact, in the preceding paragraphs, I expressed how much it meant to me that I ended up in sunny Malta, having lived more than 25 years in another country before. Living next to the sea was something I was actively searching for after I first saw the small town of Sitges in Spain, where my university sent me to present a poster at a research conference.

Even though fun and hobbies are important, bear in mind that people who hate their jobs pay too much for pursuing their hobbies: wasting 40 hours on a job you hate, pretending to be productive, was simply not the way to go in my opinion. Instead of escaping from my life for a couple of weeks per year by taking a holiday somewhere exotic, I worked toward making my own life a holiday in itself. My life became good enough so that I never wanted to escape from it.

This is the reason why I challenge you to pursue goals that mean a lot to you. You might want to become a team lead or a CTO, but ask yourself why. If you find a strong enough why, eventually, you will figure out the how.

Breaking down your goals is not easy. Depending on where you are in your career and how far you are willing to see ahead, you can target a big goal in 10 years, 5 years, or 3 years. Then travel back in time from the destination, and make sure you come up with meaningful steps.

For instance, suppose that you have just graduated from university, you have 3 months of internship behind you, and you started working as a junior developer in a software house, where you have the opportunity to contribute to various projects on a low level. Imagine that you live in Portugal and you speak relatively good English and perfect Portuguese. Imagine that your dream is to establish a life in Australia one day.

I would start with looking at the visa requirements and determining what kind of career capital I have to collect in what time span in order to make it. Furthermore, I would also start planning the financial conditions of starting a life in Australia.

Then I would determine that within 5 years, chances are that I would be able to move to Australia with a work visa, as a software development team lead. My goal would therefore be formulated in the following way: I will be a software development team lead in Australia, in a sunny city next to the sea, preferably Sydney or Brisbane.

You don't have to specify the exact city at this stage; you can clarify it later.

Your next job is to figure out what you need to reach in 3 years in order to be on a good track for your 5-year goals. Most likely, you have to be in a team lead position by then.

Therefore, your goal now is to become a team lead within 3 years, in order to collect at least 2 years of relevant experience. Notice that this is a lot stronger goal for you than just saying out of the blue that you want to be a team lead in 3 years. The motivating factor of surfing in Australia during your weekend and having a nice, comfortable life is very valuable when it comes to waking up and making a difference every single day.

If you lack soft skills and communication skills right now, you might notice that you need to acquire these skills. Therefore, make sure you increase your career capital by creating a study plan on assertive communication, nonviolent communication, active and constructive communication, and soft skills; and you might start with purchasing *How to Win Friends and Influence People*[9] by Dale Carnegie today, and read the first chapter. Your study plan is also included in your 5-year plan.

Most likely, 2 years from now, you should already be on the right track to become a team lead, knowing whether you stay with your current company or you are deliberately looking for another company, where a team lead opening is possible within a year. Worst case, you will already know what to state in your interview clearly: you are looking for a team lead role. Go out for what you want; your interviewers will back you up if you will have gathered enough career capital by the end of this 2-year interval.

Talking about career capital, I would also target writing a successful book on leadership in 2 years. It does not matter that you have zero experience in leadership. The perspective of a junior starter is very valuable, and other people will follow your example.

One year from now, you might be thinking about getting rid of your junior title and making more money.

You have to plan how much money you need to start a comfortable life in Australia and break down how much money you allocate for that. You either save money, and write down how much money you need to save in order to collect the required sum in 5 years, or you start thinking along the lines of getting other sources of income. This can be freelancing on the side, starting a small business, or creating some products in the field you specialize in. There are even wilder ideas out there. Designing your money-making side-projects will also be backed up with the big dream of increasing the quality of your life by moving to Australia.

A couple of side projects may help you on GitHub. Blogging about these side projects is also an advantage, meaning that your blog should not only be about your view on leadership but also about leading by example, by delivering professional solutions.

---

[9] Dale Carnegie, *How to Win Friends and Influence People* (Simon and Schuster, 1937).

Notice that some people only go this far. They launch a blog or start a side-project, because they know it may help them in their career in 1 year. This is not too motivating.

If you followed the preceding path, you would already know that you have to work on your blog, because you want to get rid of your junior title in 1 year and you want to become a lead in 3 years, in order to gather 2 years of experience, so that you can live the life of your dreams in Australia in 5 years.

You can then design your next year accordingly. Set yourself some milestones in 9 months and in 6 months about your blog and your side-project. In 3 months, you might want to start your side-project and develop blog content until then. In 1 month, your blog should be up and running. Within 2 weeks, you should finish studying WordPress, blogging, and technical writing; and you should already have a solid web site ready to be published.

Develop a weekly routine for yourself consisting of tasks that you will accomplish regardless of the circumstances. You may decide that you work 1 hour each week on your side projects. You may find that in your situation, writing a blog post each Thursday night makes perfect sense. Alternatively, you may set up your rules to work out four times a week to stay healthy.

Make sure you don't even think about these rules; just execute them. Once your goals change, you may adjust your weekly routine. However, debating with yourself whether you want to do something you already committed to results in depleting your willpower and drifting away from achieving your goals.

As long as your internal drive is strong enough, your weekly routine will be easy to execute. If your long-term goals don't mean anything special to you anymore, it is time to think about more meaningful goals and redesign your next 5–10 years.

We finish with the breakdown at the day of today. Ask yourself, what do you need to do right now, in order to reach your 5-year goal that's more meaningful to you than anything else?

Breaking down a 5- or 10-year plan is one of the most successful ways of lighting up some fire inside you that will give you all the energy, willpower, and motivation to keep on marching forward in the direction of your goals. This fire will support your

- Technical learning plan
- Soft skill learning plan
- Knowing what positions you are targeting
- Creating side-projects

- Financial decisions

- Health and well-being

- Lifestyle

What most people do is they buy the latest bestseller self-help book and read it pretending that they are most conscious about personal development than others.

For instance, you might have heard of the mystical power of written-down goals. Among others, Zig Ziglar illustrated it with a story about the students of the Yale University: 3% of the students had written down goals, and they outperformed the rest of the 97%.[10]

This cliché keeps on manifesting in self-help literature. Multiple authors have cited this story when talking about goals. Yet, the research seems to have been a mere urban legend, as according to the best of my knowledge, it never existed.

Goal setting is not about the magical power of writing down your goals to make them happen. Goal setting is rather about making the reasons behind your actions more conscious. There should be continuous and consistent action taken in the direction of your goals; otherwise, you would just join the millions of dreamers following the advice of *The Secret* and hoping that written-down goals would one day manifest in front of their eyes.

Even if the study about the 3% with written goals was true, the cause and effect relationship would still not be there. It is not the piece of paper that manifests your goals, but rather your commitment to taking action that is necessary to reach your goals, while not focusing on anything that just wastes your time.

If you stay committed for 5 or 10 years, you will be able to achieve miracles. If you only set a goal for 1 year, chances are you will overestimate your results massively.

Don't forget rule 4 of *So Good They Can't Ignore You*. Think small; act big! Focus on the actions that matter in the moment; don't get caught up with solving problems, where more career capital is required.

Be flexible in your approach! Chances are as you gain career capital and you get to the cutting edge of your current field, new opportunities will become visible to you. In this case, feel free to spend some time on changing your goals and designing a new future for yourself.

---

[10] Yale confirmed in 2017 that no such study exists according to their knowledge. Source: https://ask.library.yale.edu/faq/175224

For instance, maybe you figure out that someone contacts you 1 year after you start writing your blog, asking you to join their company as a team lead. Provided that you have acted as a lead even in your junior years, with none of the powers, but all the responsibility, and the ideas you blogged about are already cutting edge, you will have enough career capital to become a lead in 1 year. This scenario is very unlikely, but it is not impossible.

My last advice about goal setting is to enjoy the path leading toward success. Because happiness comes from progress. Getting the goal does not make you happy for a long period of time. If you need common-sense coaching of being in the moment and enjoying the path leading toward your goals, I can highly recommend reading the book *Way of the Peaceful Warrior: A Book That Changes Lives*[11] by Dan Millman.

If you make sure you enjoy your path and move toward your destination at a sustainable rate, you will maximize your chances of mastering your career.

Exercises

1. Plan your next 5 years based on the method described in this section.

2. Create your weekly schedule. Make sure it's a schedule you enjoy for a long period of time. Don't burn yourself out!

3. Take the first action today in the direction of your destination.

Summary

There are many career paths in software development. You can become a generalist, a specialist, a T-shaped professional, a leader, a manager, a domain expert, or a process expert.

As your skills develop, you will go through several career transitions. These transitions assume that you take more responsibility to function in more senior positions.

To navigate between these positions, you need to know where you are now and where you want to be. Plan your path backward from the end state, and execute swiftly so that you not only reach your goals, but you also make the journey exciting.

---

[11] Dan Millman, *Way of the Peaceful Warrior: A Book That Changes Lives*, revised edition (HJ Kramer, 2006).

Your Online and Offline Presence

Personal Branding for Software Developers

Job satisfaction is an important topic for software developers. Some developers just want to earn more money. Others want better job security or better working conditions. Other developers want to leave a heritage by affecting the lives of many people and building something they are really proud of.

In order to continue your journey in the direction of your destination, it is inevitable to formulate your message and communicate it to the outside world. This way, other people will find out about your career goals, and you will get some allies on your journey. By building your online and offline presence, you will handle potential excuses that could prevent other people from cooperating with you. Others will find out what makes you special. Some people will conclude that it is worth for them to cooperate with you as an expert.

© Zsolt Nagy 2019
Z. Nagy, *Soft Skills to Advance Your Developer Career*,
https://doi.org/10.1007/978-1-4842-5092-1_4

In order to accelerate your progress, you not only have to learn some skills, but you also have to invest in making those skills visible.

We will start with designing your personal brand and explore what you want to be known about. This decision has a direct influence in crafting your social media content, projects, and blog content.

After crafting your message, we will start building your social media presence so that other people will want to work with you.

Opposed to many professions, in software engineering, demonstrating what you know is more important in getting you hired than just talking about what you know. Therefore, we will discuss the fundamentals of building a portfolio.

After talking about the principles behind creating a successful portfolio, we will discuss how a blog can be one of your best assets in building your portfolio. Blogs are interesting for multiple reasons. First of all, writing organizes your thoughts better; second, you will challenge yourself to learn and share new things; and third, your efforts will establish you as an expert of a certain field.

Creating webinars and YouTube channels takes everything to the next level, because talking about a topic live in a knowledgeable way is even harder.

We will also identify time wasters along the way. Based on the Pareto principle, 80% of your efforts will come from 20% of your tasks. Although this sounds like a cliché, it highlights the fact that many of the tasks we deal with don't matter that much. We will identify obvious time wasters that don't get you ahead.

We will investigate the role of meetups and conferences. Oftentimes, you will meet people you can cooperate with, and you can hear brilliant ideas.

Let's get started with your personal brand.

Design Your Personal Brand

Goals of this section:

- Discover your values.

- Determine the value you provide as a developer.

- Learn how to create a personal brand for your online presence.

At first sight, you may think that personal branding has not much to do with software development. In this section, you will figure out that it is not only necessary, but it is also a trending topic in the 21st century.

As social media has become popular, we can connect with more people than ever before. There is an increasing number of people trying to grasp our attention. Therefore, attention has become a more scarce resource than ever before. Others won't listen to you if it does not give them the perception of a positive return on investment.

People direct their attention toward valuable information. Therefore, we will introduce the concept of personal branding based on our values. While working on our personal brand, we will make our values clear to others.

Your values can be used to shape an image of the type of professional a certain segment of companies want to hire. This means you want to help a special segment of companies find out more about you. If you know what types of companies you want to work for, you will waste less time with the job hunt. One way to know what type of companies you want to work with is to find out more about your values.

The key to personal branding is the demonstration of your values. The opinion of your audience is like a lens of a camera. The lens can be distorted, it can be broken, or it can also be dirty. We have absolutely no influence in shaping the camera. We can only influence what we show from ourselves. This is what personal branding is all about.

While you cannot influence the camera, you can influence

- Which lenses to target
- How you stand in front of the camera

Before standing in front of a camera, think about the professional skills you need to get the position you are looking for, and shape your personal brand accordingly. Create a lasting impression by crafting a story and transmitting emotions.

What Is a Brand?

A brand is a story with some expectations. When creating a brand, you create a promise others will expect you to deliver. Some expectations are associated with higher value, while other expectations are associated with lower value.

Expectations are important because people rely on what you have to offer. The word market in the job market means that the laws of demand and supply hold. If you don't focus on personal branding, the job market won't know much about you, resulting in low demand for your services. If you send mixed signals to your potential employers, there will be an aura of uncertainty about you.

This is why it is important to identify information worth broadcasting about yourself.

There are multiple methods and frameworks for developing your personal brand. I can highly recommend the Personal Branding Canvas by Luigi Centenaro.[1] If you go through this canvas from the perspective of your career, you will draw very useful conclusions that will eliminate some uncertainties in the eyes of people employing you in the present or in the future.

The 21st century is all about branding. Brand owners need their brands to create a maintainable relationship with their customers.

In order to create a credible brand, you have to know where you are right now and where you want to be in the future. If you are working on the materials of this book sequentially, you have already bridged the gap.

In order to support your journey to the desired destination, you will need to know your values.

Determine Your Values

Most people identify themselves with their current jobs. Bear in mind that your job is a part of your personal brand, but there are many other things you should be thinking about.

Your everyday actions are based on your values. Find out what your values are, so that you can get closer to who you really are. You might already know what your personal values are. If you need help, fill out the VIA Character Strengths survey.[2]

Half a year before filling out this survey for the first time, I determined my values on my own, trying to order the character strengths I found in an article about Tony Robbins.[3]

Back then, I had a lot of challenges in my life, and those challenges came from trying to meet expectations of others. My values were distorted, because they were not about what I truly wanted; they were about what others wanted from me. Whenever I had a moral dilemma, I thought about my values and acted according to them. These actions were not comfortable at all. I felt that something was missing.

After filling out the VIA Character Strengths survey, everything made sense. The top two values I had thought characterized me did not even make it to the top ten in the new list. This made me question my values, whether they were mine or they were just there to meet the expectations of other people.

---

[1] www.personalbrandingcanvas.com
[2] www.viacharacter.org/www/Character-Strengths-Survey
[3] http://sourcesofinsight.com/change-your-values-change-your-life/

If your values are not yours, you live in denial. This is why it is important to choose your values based on what you want and not based on who other people want you to be.

Following the discovery of my character strengths, everything made sense. My decisions were comfortable, and I even found out why I loved the company I worked for back then. It also made sense why I got promoted so quickly. The top four values of my company were included in my own top six personal values. Once you know that not only your company but also you live your values, the cooperation becomes very smooth.

Oftentimes, our real values are different than our perceived values. One reason is the impact of the society. Many people feel guilty when they want to pursue their own dreams against the expectations of society. Don't go down this path! You will want to live your dream, and not someone else's dream.

You may now rethink your desired destination. Feel free to go back to Chapter 3, and plan your next 5 years if you were constructing someone else's dream. Goal setting is an iterative process. Don't be afraid of changing your goals as you discover new things about life or new things about yourself.

■ **Note** Another aspect of living our values is that people tend to defy their values to meet their needs. This is why Chapter 2 is so important. A strong self-image, high self-esteem, growth mindset, and general lack of victimhood contribute to your ability to meet your needs. This in turn strengthens your ability to live your values without distractions.

Notice the layers of our approach:

- Chapter 2: Learn how to meet your needs from within.
- Chapter 3: Know your vision and purpose.
- Chapter 4: Discover your values and convey them online.

One of the most important elements of your personal brand is your perceived performance. You need to show that you utilize your strengths for the proper causes.

Your performance has to be directed by your values and your moral code. Performance without a moral code makes no sense, and it is also demotivating. For instance, when I was thinking about my own moral code, I came up with the following items:

- I commit to excellence whenever I deliver a solution.
- I commit to creating maximum value possible.

- I have clearly defined quality standards, and I refuse to lower my quality standards in the long run.

- In the short run, if business needs require me to compromise quality standards, I cooperate and ask for time to clean things up properly afterward.

- I cooperate with everyone in an environment based on trust and respect. I refuse to cooperate with tyrants.

- I train myself every single week and unlock new abilities. This results in exponential growth in my skill level.

- The act of problem solving matters to me more than lexical knowledge.

- I am flexible in my approach. Common sense may override any rule I have in life and in my career.

- I celebrate good character traits. I reject toxic behavior at the workplace and defend my boundaries.

If you want to go the extra mile, I can highly recommend reading or listening to the book *Principles: Life and Work* by Ray Dalio.[4] I know, Ray Dalio is not a software developer. However, his principles inspired me in my own career, because as a manager and as a meritocratic leader, he inspired me to redefine my own view on professionalism.

I hardly ever communicate these rules. These are my way of delivering results, and people have a very good idea about my moral code, as my behavior is **consistent** with them.

For instance, when my manager approached me that we would lose a big client unless we got something done in 2 weeks, I told him I would do everything I can to deliver in 2 weeks and he can count on us. However, I also needed up to 2 weeks to clean the mess up in the process. I got the approval of the extra 2 weeks immediately, and I could count on the extra 2 weeks of refactoring time.

In another project, a change in specification surfaced, resulting in at least 2 days of extra work. My opinion was that following our agreements, the change should be rejected. We can only deliver quality if we respect the scope of the milestone we are dealing with. I still encouraged the discussion to take place, as an incremental update right after the release of the milestone would contain these changes.

---

[4] Ray Dalio, *Principles: Life and Work* (Simon & Schuster, 2017).

Common sense overrides these rules. For instance, if a change in specifications resulted in getting rid of 3 days of work, it is evident that we should make an exception and change the scope.

Whenever you have to act, think about your values, and act according to your rules consistently.

The Value You Provide

Determine what you have to offer. If you lack experience and you offer WordPress development services, you know that there is a lot of competition. The value you provide is most likely that you are cheap.

If you have a shiny portfolio as a WordPress developer, you can expect that you get more offers from employers, but there are still many people out there delivering work of similar quality.

A developer known for creating award-winning WordPress sites is a different category. Also, if you state that you specialize in implementing fast headless CMS solutions and you have a tagline "Award-Winning Headless CMS Developer in WordPress" and you attach the latest version number, you can be assured that people will find you.

Note   If you keep updating the version numbers of the software you are using, people will know you are fully up-to-date. There is nothing worse than a resume containing outdated skills such as PHP5 or ECMAScript 5. Even AngularJs is a trap, because from version 2.0, this framework is called Angular.

To take things further, if you are one of the contributors of a headless WordPress open source repository, you should hardly have any problem finding a job or even becoming self-employed.

Think of all the tech skills you want to be known for, and don't forget to specialize. I have seen so many resumes claiming that their owners know JavaScript, PHP, MySQL, .NET, Java, C, C++, Prolog, Haskell, Object Oriented Programming, UML, Functional Programming, different frameworks, and even Microsoft Office. Applying some self-criticism on myself, one of these resumes I saw was mine just before I graduated from university. Back then, I had really no clue about how to market myself, and only my past achievements sold my services. Pick one niche and stick to it. If it is Angular 7, then delete every other irrelevant fact from your resume. Who cares that you have a driver's license? Who cares that you know Microsoft Excel? Are you a software engineer or an administrator in finance? This advice applies for all traces of your online and offline presence. You don't have to hide your past working experience, but don't display a page-long boring list of irrelevant skills.

Your soft skills also count. Determine what you want to be known for, and make sure you keep on practicing those skills. Make sure you consistently apply them. You should not write down in LinkedIn that you are an excellent communicator. Instead, create a compelling mission statement, edit your job descriptions, and use social proof, when testimonials of other people clearly state that you are a good communicator.

According to Luigi Centenaro, you can offer four types of benefits.

Functional benefits mainly depend on your regular skills and your problem-solving ability. If your employer knows that you are capable of solving a problem, you have a good chance of offering these functional benefits. For instance, the following is a potential functional benefit: leading a team of three developers to develop a client-side rich web application for the purpose of creating a user interface to a service the company is offering based on a REST API.

Emotional benefits are mainly based on soft skills. You need to create a lot of value, as no soft skills substitute value. On the other hand, a genius who is very hard to talk to, often comes in drunk, does not shower, and sometimes refuses to work is not wanted anywhere either. Everyone has to offer the basic benefit of being a reliable person, requiring low maintenance. Leaders, and people communicating with stakeholders, have to offer a bit more than that.

Showcasing emotional benefits is hard, because you never want to brag about how awesome your soft skills are. Use storytelling to develop narrative influence, make sure other people know you for your consistently applied soft skills, and meet the needs of others.

The **self-expressive** benefit group is a bit harder to express than the first two. It is mostly about delivering the promises of your personal brand, delivering the promises of what you want to be known for.

Social benefits are more obvious, and they are somewhat linked to emotional benefits. A person with good soft skills is likely to meet both. However, social benefits go beyond being able to communicate. Other people should relate to you better. This means your character should be mentally mature. It also helps if you have an interesting life and hobbies on the side. You may be known for hosting a meetup for core training every weekend. Alternatively, you may be the person introducing half-hour-long meditation sessions in the company to boost efficiency. You may be the person establishing a book club or a running club in your company.

Your credibility comes from

- Consistency
- Showing what you are capable of in the context of what you have to offer

- Focusing your profile by not showing irrelevant details
- Creating value that fulfills the needs of your employers or clients

Mission Statement, Vision Statement, Tagline

Once you know your rules, it is easier to create your vision statement, your mission statement, and your tagline for your online presence. A lack of these statements implies that other people think about you as someone replicable.

A **mission statement** answers how you will get there. These are things you should already be doing, or things you want the world know you are doing.

A **vision statement** answers the question what you are heading toward and details what kind of person you want to become. Do you remember the HR cliché question "How do you see yourself in 5 years"? A vision statement is a short summary of your thoughts.

A **tagline** is a very short summary of the core of what you are doing. Try to use just a few words.

Note If you have a hard time coming up with a vision, a mission, and a tagline, think about your values. Think about **why** you want to make a difference in the world. What drives you? Then think about **how** you are going to do it. Finally, think about **what** you are going to do. For inspiration, check out Simon Sinek's talk titled *How Great Leaders Inspire Action*.[5]

Use your vision statement, your mission statement, and your tagline in your LinkedIn profile, your resume, and be prepared to communicate it verbally.

Your mission statement, vision statement, and tagline should communicate why you are the most suitable person for the projects you undertake. Apply this message consistently online and offline as well, let it be LinkedIn, Twitter, your blog, or even your Facebook account.

For instance, suppose your **vision** is to create an environment where collaboration brings professionals together regardless of their specialization. Why? Because you love working with people who contribute instead of competing. Because deep down, it frustrates you that you see a lot of precious effort wasted on a regular basis, just because developers, UX, product management, and business don't understand each other due to the lack of common language.

---

[5] www.youtube.com/watch?v=qp0HIF3SfI4

Note When explaining why you feel strongly about your vision, the easiest answer is "because I love to make a difference this way." Your emotions are universally true statements. If you state that you love something, it is accepted as a justification by default. It also shows that you are motivated beyond earning a salary for your work.

The **mission** statement deals with the how question. For instance, you can make a difference in the lives of collaborating people by bringing UX and frontend development teams together by defining ways of collaboration that help build a component library. This library helps developers understand UX concepts and considerations and helps the UX team understand how developers think and use their components.

Your **tagline** could sound as follows: I bring UX designers and web developers together by building maintainable, easy-to-use web component libraries.

Your brand fails to fulfill its primary purpose whenever you fail to communicate what other people expect from you based on your past promises. As you are the one creating and shaping these expectations, every single message of yours is important. For instance, if your personal brand shows that you are a person to be taken seriously, publicly posting an image on Facebook in a situation most people find shameful rarely makes sense. Other people get away with it, because their personal brand is not in conflict with having fun in their way.

Some companies do not tolerate this behavior from anyone. Other companies may reject you as soon as they look at your Facebook profile. Different cameras construct different images of you. I suggest showing them your professional side, and make sensitive content on Facebook only visible to your friends.

It always makes sense to be yourself. Creating a fake persona you pretend to become, but you don't want becoming, makes no sense. If you can't impersonate this persona, others will figure it out quickly, meaning that your job may be at stake by the end of your probation period.

If you would like to take these ideas even further, create an **elevator pitch**. An elevator pitch is a couple of sentences describing what you do for a living, when you meet a stranger. It should be very interesting; otherwise, people won't remember anything about you. It should also be understandable, even if the person you meet has no idea about the field of your expertise. Using just a couple of sentences, express yourself in the best possible way.

Headshot, Fashion, Grooming

Once my manager made an observation that some software developers only shave and dress well when they are planning to leave the company and start interviewing elsewhere. I reject this generalization, but unfortunately, I can recall at least five software developers who showed exactly this tendency. By reading this section, you get a chance to stand out.

Your appearance is not everything, but it counts. Your wardrobe, your personal style, is also part of your personal brand. In some companies, dress code is mandatory. If you try to slightly dress above the dress code, you cannot go wrong.

In some startups, suiting up would rather work against you and would create the impression that you are not one of them. You can still wear business casual clothes if you feel comfortable in them.

When I was a consultant, my agent asked me to wear business casual clothing. Even though other people were dressed as casual, I stood out in a positive way, I was respected, and some people even commented that there was finally someone who dressed up in style. The key to receiving these comments is that your clothes should feel comfortable. It is not enough to wear the right clothes. You also have to know how to wear them.

Fit is generally more important than the brand you are wearing. I admit, back in the days, some of my clothes were a bit loose on me, especially after I started losing weight. Once I learned how to pay attention to my clothing and bought better fitting clothes from the same brand as the one I had used before, some of my colleagues started complimenting me. Once I had lunch with a female colleague of mine, and she confessed that the whole department noticed that I became more stylish.

Why does this matter?

The reason why people judge you positively based on you clothing well is the **halo effect**. Their impression in one area of your life influences their opinion in another area. In other words, if you dress well, more people are likely to think that you also work well. The only exception is when you are not comfortable in your clothes or your clothing is in full contrast with everyone else. In these cases, people are not impressed, and there is nothing positive that can influence their view on your work.

Normally your results speak for themselves. However, you never have a second chance to make a first impression. Clothing and the halo effect are a big part of first impression.

Your clothing should be consistent with your message. If you want to be an approachable, down-to-earth person, it is absolutely fine to go to a conference and present in a leather jacket, jeans, and a T-shirt.

■ **Note**　As a software developer, I do my best to think as rationally as possible. Unfortunately, rational thinking in terms of fashion is hard to master. The key to understand the rationale behind fashion is to accept that wearing fashionable clothing bears a cost. Fashionable clothes are sometimes uncomfortable. Some accessories have no function, and therefore they make no rational sense. In an evolutionary sense, the reason why these items appear fashionable is that the bearer of the item can afford the cost of wearing the item. Think about the time and money needed to iron your shirts so that they appear flawless. It is not the features of an item that make it fashionable. It's the message that you are comfortable with wearing it.

Think of clothing as a way of expressing yourself. Do you want to be known about being an armchair athlete, because a funny T-shirt of yours says so? It is all right, if humor is one of your values, as you are expected to wear funny T-shirts every single day. Otherwise, it just adds noise to who you really are.

Once you are comfortable with your clothing, have a headshot taken, and use that headshot consistently everywhere. Do you want to be approachable? Wear casual clothes. Do you want to display high status? Wear a suit that fits you. Do you want to look like a clown? Wear a suit that does not fit you at all, wear a tie with a clumsy knot, make sure your shirt is creasy, and make sure your colors don't fit.

Reach Your Audience

Your brand is worthless if your target audience does not reach you. For instance, if you are changing careers, your new personal brand will not be too valuable for your current employer. Writing a blog and launching some relevant side projects help. However, the target audience will have to find you.

Get your name out there! If needed, answer Stack Overflow questions, go to Quora, write on medium.com, go to meetups and conferences, present in meetups, create podcasts, create YouTube videos, create something meaningful on GitHub, or just go to job interviews.

Find role models in the software industry, and check what they are doing. Maybe you can learn something yourself too. Do not imitate their personality, just check for ideas, and think about them in the context of your own exposure.

If you deliver a lot of value, people will start talking about you, and your exposure will be increased on a natural basis. Meeting new people still increases your chances, as you can choose from more opportunities. Therefore, in the upcoming chapters, we will work on increasing your exposure.

Exercises

1. Determine your own values based on the VIA questionnaire mentioned in this section. Feel free to override these values if they seem wrong. Start resolving inner conflicts.

2. Develop your business code of conduct that defines how you act as a software developer. I highly recommend reviewing Ray Dalio's book called *Principles: Life and Work*.

3. Think about the value you have to offer, including all four benefit types defined by Luigi Centenaro.

4. Formulate your vision, your mission, and create a tagline.

5. Create an elevator pitch, and practice it in front of the mirror.

6. Identify your current personal style, and consider if you need an upgrade. If yes, take one step today toward improving your ability to create better first impressions.

The Irresistible Resume and Cover Letter

Goals of this section:

- Use your personal brand to determine the content of your resume.

- Make your resume and cover letter authentic.

- Understand what employers want and address your application package to them.

- Enrich the reader's experience with storytelling.

- Connect your cover letter and resume with hook points.

- When the position justifies it, consider going the extra mile by creating a video.

It is time to transform your personal brand into a resume.

I understand writing resumes may not be your favorite activity. I understand that you may think you are writing a piece of paper that will not even be read. Many people may close your resume after 5 seconds. Therefore, we will have to generate some leverage first.

Imagine creating a resume that will grasp the attention of many people. Imagine how easier your life is going to be during the interview process if your resume places some clues toward what you are going to talk about. The point of a resume is not an entry ticket to the interview. The real point is to set the frame of the interview and make the whole interview process easier.

We shift the focus from getting an entry ticket to setting the frame in the interview. After all, software developers are in high demand. Some companies will interview you even with average resumes. The question is: How will you make your life easier during the interview? How will you maximize your earning ability?

You will also save time by learning how to create a good resume. It is a small time investment, but you will get more opportunities, and it will take you less and less time to create your resume as you get hands-on practice. As you know from the last chapter, marketing yourself is an inevitable skill today. Learning how to write good resumes may earn you hundreds of thousands of dollars if not a seven-figure sum throughout the course of your career.

Your resume is not a list of previous jobs. It is a demonstration of how you live your life as a professional, clearly stating where your career is heading toward.

At the time of writing this section, I have had more than 9 years of active experience in evaluating job applications. Even though we don't get hundreds of resumes in 1 week, I still had difficulty with filtering applications. Having seen more than 500 applicants in action, I concluded what many people do wrong and just a couple of people do right when writing a resume.

Most people create an enumeration of their employment history, their studies, add some clichés and leave the rest to chance. When chance never comes, they start spreading their experience that the job market is not working, and you should be happy if you have a job. All you need to do to escape from this crisis is to elevate yourself from the bottom where the masses compete and target the top, where the best professionals reap the rewards.

As we concluded in the last section, every person, every hiring manager, or every company interprets your personal brand differently. Your personal brand manifests as different images captured by different cameras. Therefore, our personal brand is elastic by nature.

Elasticity does not mean that we want to meet the expectations of other people by denying who we really are. It just means that we will emphasize different characteristics of our identity, our skill set, and our experience based on the position and company we are targeting. This elasticity will be clearly visible in your resume and in your cover letter.

This is the exact reason why professional resume writers don't solve all your problems. Regardless of the quality of the resume you get, you will have to tailor it. The professionally written resume will be a great asset for the one position you applied for, but later, you will have to make modifications, and your modifications will have to be consistent with the style of the resume.

Consistency also implies that your personal style should be consistent with the style of the resume. Based on the four essential soft skills we value in this book, taking the responsibility of writing your own resume will be a valuable skill for you. You will reap the rewards of writing your resume during the interview, as more neural connections will form in your brain.

Think about it for a moment: your resume is one marketing tool out of many. In the 21st century, marketing yourself will be essential. Unless you employ your marketing team, you will have to get better in marketing yourself. The first step is taking charge of your resume.

Regardless of whether you would like to spend $100–$1000 on a professional resume or you write it yourself, keep in mind that it has to be *authentic*. There are many deceptive job applicants out there that it has made hiring managers less forgiving than ever before. Authenticity is a huge value. In this society, people do everything they possibly can to change their image based on the expectations of others, up to an extent where people lose their identity.

Bear in mind that if you are not willing to learn this skill, you will still learn a lot from a professional resume writer. Unless you purchase a cheap, second-rate service, a professionally written resume will look better than a badly written resume. However, sooner or later, it is worth learning how to market yourself on your own.

If you choose to hire a professional resume writer, do your best to extract as much knowledge about the process of writing your resume as possible. Apply lateral thinking, and figure out where you can learn your skills. Then polish your LinkedIn profile. Polish your blog.

If you cooperate with a professional resume writer well enough, your resume will be authentic. Therefore, I am skeptical when someone offers me to write a resume for $100. There is just not enough time to do the iterations and establish a common ground.

What Are Companies Looking For?

First of all, companies want to make sure you are competent enough to contribute autonomously. Many companies have been tricked by people who seemed to know what they were doing, but they created more trouble than value. This is why you have to demonstrate a high level of maturity in your application package and provide some proof.

Companies are also looking for employees who do not need to be managed too much. Companies need people who can solve meaningful problems on their own and suggest solutions instead of feeling blocked because of small obstacles.

Companies want people who fit in their team on a cultural level. We all know how hard it is to work with people who take your lunch from the fridge, refuse to work on their tasks, or keep distracting the attention of other people.

Beyond the basic needs, I can refer you to the book *Influence: The Psychology of Persuasion* by Professor Cialdini.[6] Your resume and your cover letter should communicate the six fundamental principles of influence.

Cialdini's six principles, Reciprocity, Consistency, Social Proof, Authority, Liking, and Scarcity, have been translated for the workplace. For more details, check out the article and video titled "Principles of Persuasion".[7]

Creating a Product-Market Fit

Think about your career as a business, where you offer your services as a product. You have to find a good place for these services in the job market. Your job is to create a fit for your services in the market.

In order to create the product-market fit, you need a resume, a cover letter, and the interviews; you need to set your price. In the background, all other manifestations of your personal brand support finding the product-market fit.

If you want to target everyone, not many people will bid for your services. This is why it makes absolutely no sense to enumerate all your past employers, all your skills, and rely on just data. Yet, most people use exactly this strategy, and wonder why there is no market for their product.

Many people don't know what to reveal about themselves. They create a data-based resume and a cover letter full of clichés. All you need to do instead is to demonstrate that you have the skills required for a given position.

In order to demonstrate your skills, you need to be able to tell stories both verbally and in writing. Stories create a lasting impression, and they don't sound like bragging. Stories are also assertive, because you let your audience shape their conclusion instead of telling them directly that you are a perfect asset for the company. In order to shape the proper image, you have to tell the right stories.

The interest of your employer is to find out if you will motivate your colleagues. They want to know if you will be a great fit in their team, or you will rather cost them a lot of money with your mood, attitude, and destructive behavior. You can be the best professional in the world, but if five out of five colleagues resign because of you, it is not worth for any company to employ you.

---

[6] Robert B. Cialdini, *Influence: The Psychology of Persuasion*, revised edition (Harper Business, 2006).
[7] www.influenceatwork.com/principles-of-persuasion/

Your resume should create a positive impression of your soft skills. The only credible way to do this is through storytelling. If you write clichés such as

- You are a great asset to the company

- You deliver quality work

- You have excellent communication skills

your hiring manager may easily label these sentences as nonsense. If you tell a short story about a challenge you have overcome, saving a hundred thousand dollars on detecting a communication mistake, or executing a time-critical update ahead of schedule, your hiring manager will most likely think that you will be a great asset, you deliver quality work, and you are dependable to deliver excellent results.

Furthermore, human fantasy tends to work in your favor by associating even more good character traits with you thanks to the previously mentioned halo effect.

Make sure your employer will start narrating your resume in such a way that you have a massive profit-earning potential for their company. This will increase the salary expectations that will likely succeed with your potential employer.

Letting Go: Storytelling Without Limitations

Storytelling is very hard, especially if you are a beginner. If you are not used to writing, you will face some challenges. In a professional resume, the length restrictions will also act against you. This is why you need to learn about the fundamentals of storytelling.

When I wrote my first article on zsoltnagy.eu, I was afraid of publishing the article on Hacker News. I thought people would criticize me. You may have the same fears when thinking about rewriting your fact-based resume into the format I am suggesting.

You may also feel that you are not worthy of mentioning an accomplishment of yours, or even to target a lucrative position. This is another limiting belief you should eliminate. Believe me, doing your best and writing the best resume you can is worth every effort. You may get your dream job with less experience than the most suitable candidate.

Your story will be authentic and credible, if you first work on yourself and let go of your fears. You will open up, you will become vulnerable, and other people will relate to you better. This is how you create a lasting impression.

People love stories. Stories grasp our attention and make our message unique and relatable. A good resume tells your story in a relatable way. At the end of reading the story, the reader will know about you, about your journey, and is more likely to cooperate with you in your quest.

The secret about creating a great story is to analyze your audience. Your resume is not about you. Your future employers will be interested in the value they gain by selecting you. Ask yourself what skills your potential employers are looking for. What problems are they facing with? What are your employers afraid of while evaluating your resume and thinking about employing you? Always personalize your story based on the job ad. Use their expressions, and show them that you didn't just randomly send out hundreds of applications with the exact same content.

Let go of all the conventions. There are absolutely no conventions when it comes to resumes. The worst thing you can possibly do is to use a popular template, such as the Europass resume format.

A story is about a challenge.

The Story-Based Resume

Let's lay down the foundations for creating a resume.

It makes sense to start conventionally. If you choose to include a photo of yourself, use a professional headshot. Insert your personal details. A title is not needed for your resume, as it's redundant. Displaying Resume or Curriculum Vitae at the top of page I makes no sense; it just makes you less relatable.

The top section of your resume should contain links to your online presence. Link to your LinkedIn profile, your GitHub profile, your Stack Overflow profile, a Medium account, and your blog if you have any of these. If you have done anything else worth mentioning, consider linking information about it.

The first section of your resume should be about your professional goals, your motivation in what difference you would like to create. Think about your vision, your mission statement, and your tagline. Formulate it in a way that it looks good with your employer. Most people never do this. Therefore, your hiring manager is not likely to read this section, and can't wait to find out more about you.

In the previous section, you constructed your elevator pitch. Construct a story around some of your accomplishments in a similar fashion. Back up your accomplishments with data, and construct a story around it.

The skeleton of the story is the following:

- What was the goal of your project?
- What challenges did you face?
- How did you overcome these challenges?
- What results did you get?

For instance:

- Coordination of the development of a user interface aimed at automatizing the work of account managers. Due to continuous cooperation with account managers, we managed to understand and automate a large chunk of their work. This allowed us to scale from 100 to 800 clients without hiring anyone, saving several hundred thousand euros per year.

- I was responsible for optimizing the performance of our endpoints so that our average response time is decreased from 2.5 seconds to 0.4 seconds. The challenge was that the queries were constructed out of thousands of SQL statements, and my objective was to design a query planner that eliminated 90% of the queries. As I got the time constraint of 2 weeks, I had to focus on changes that made the biggest difference. The end result correlated with increasing our client retention rate. Our server load was decreased by 10%.

- When I joined company X, I was given the task of reducing our emergency hotfix ratio from eight hotfixes a day to one per week. This made me think out of the box, as we had to measure whether changing the development process, reducing the technological debt, or encouraging knowledge transfer, the lack of automated testing, or the way how the team had treated hotfixes made the most significant difference in reducing the error rate. I love these challenges, as they are not only technological but also involve people. The metrics motivated me, and after 18 months, it seems that we have reached our target thanks to the help and commitment of the whole team.

Make sure your stories are not filled with numbers and foreign words. This story is not optimal: Replacing the Backbone+Marionette stack and the Stickit bidirectional model-view binder with React and Redux in a 150 kloc code in 28 business days. The task included rewriting 2735 Mocha+Chai+SinonJs unit and integration tests.

Following your top accomplishments, you can tell your reader about your employment history; but instead of focusing on your position, write about your roles and responsibilities in a short story format. In order to make your story more compelling, link your responsibilities with accomplishments. If you can quantify how much money you earned your employer, or how much money you saved, and this number is not a secret, include it in your resume.

If not, think about a KPI (Key Performance Indicator) that is good enough to describe your accomplishment.

If your employment history is too long and you have changed specializations, it is acceptable to just write about your **relevant employment history** and place a link to your LinkedIn profile for more details.

Your education should receive less emphasis in your resume, unless you are a fresh graduate. If you don't have enough employment history, your personal brand and your portfolio will be extremely valuable. You will be able to showcase your side projects by telling your motivation behind creating them. Also mention any noteworthy scientific competition or activity in your college or university that may be relevant for your employer.

Format your stories and accomplishments by highlighting the skills that your employers need. Instead of creating a skills section with a long list of programming languages, it is enough to highlight that you have used these skills in practice. This does not only save you space, but you will also be less likely to be labeled as a generalist. If you insist in having a skills section in your resume, name it **Relevant Skills**, and only enumerate skills that your employer may need. For instance, if your employer needs JavaScript developers with React experience, you may mention that you are experienced with Redux, and an automated testing knowledge, using Mocha, Chai, and SinonJs. Your Java knowledge is irrelevant. Knowledge of the Zend framework is irrelevant.

Writing a good resume is a creative process that goes beyond the scope of this book. The more responsible position you target, the more money is at stake when it comes to positioning your services properly. If you want help with this process, you're welcome to contact me for a free initial consultation session at http://devcareermastery.com/resume-service/.

What Should I Write in a Cover Letter?

This is where a professional resume writer will only help you by giving you a backbone and maybe some samples. You will be on your own when writing a cover letter, and it has to be tailored based on the information you find out about the company you are applying for.

Make the cover letter personal. Read the job application, and create an authentic letter that contains just enough formalities for the position. It is possible to be informal in the case of some startups. It makes sense to be more formal in corporate environments.

Avoid clichés like "I will be a valuable asset" or "I am a hardworking, professional, highly motivated individual." These statements count as **explicit influence**. If you tell someone what they should observe, they will less likely to believe it than their own internal narration. This is why your best option is to go for **narrative (implicit) influence**. When telling a story, your future employers should conclude good things about you without you writing these things down in your resume.

One common structure is the "your requirements–my results" comparison, where you match the requirements with some accomplishments of yours. This is a straightforward and helpful format. One advice to improve this format is to place hook points in the form of references to the content of your resume.

Always talk about the why behind your application. Why them? Why now?

According to your vision, you are heading from point A (where you are now) to point B (where you want to be). A logical next step for you is to utilize your skills in a professional environment, where you can contribute the most by putting your experience in industry X into practice, working with technologies Y. This is why you really appreciated reading the blog of the company and finding out about their expertise in industry X. Refer to specific articles. You also appreciated the professional article of the CTO on medium. com about technology Y, showing that you will be able to exchange information about best practices on a regular basis. Tell them that it is important for you to work with individuals with a professional attitude and high intrinsic motivation.

This way, you show that you **qualify your employers based on your own criteria** in one compact paragraph. You also reveal a bit about your professional code and your vision, making yourself interesting and relatable. You also develop some narrative influence about your skills in the eyes of your future employer.

Being needy for the job will likely put you at a disadvantage. Don't write that you want the job badly. You have to show interest in the company, but praising the company far too much comes across as tryhard and a bit desperate. This is one reason why it makes sense to praise the company by qualifying them.

Make sure you don't talk about yourself too much in the cover letter. Even when telling a bit about your vision and your goals, match it with the company's interest. A rule of thumb on cover letters is to check for the word "I" in your cover letter, and if there are too many occurrences in a paragraph, consider rewriting it.

If you want to really nail your cover letter, give them one or two hints how you will make profit for the company. Connect one of the top three accomplishments in your resume with the cover letter, and deliver a hook point. Write down that you have contributed to a project delivering results Z, and based on the profile of the company, you are interested in contributing projects of similar or higher magnitude. Then mention an interesting fact about the project, and conclude the paragraph by referring to your resume for more information.

The Extra Mile Most Applicants Never Go

If you are already in a senior position and you are targeting a high salary, possibly in the six-figure range, it makes sense to record a nice video about yourself.

Only do this once you are comfortable. You know when you are comfortable if the video looks and sounds like your real professional communication with your colleagues. Once you are not self-conscious while talking in front of a camera, you will reach the quality required for this video. All you need to do is give a short video introduction about yourself.

If you are afraid of talking in front of a camera, I encourage you to do it even more, but only send the results in your application package, once it's good enough according to your standards. Talking in front of a camera exercises skills and eliminates fears that will be vital in your career. These are leadership skills, public speaking, and fear of rejection.

You can start small by recording yourself on a daily basis. First, talk for 1 minute, then for 2 minutes, and then for 5. The topic is arbitrary. Then rewatch yourself. This process is called **stream of consciousness**. Initially, it will be an uncomfortable feeling to watch yourself, but quoting Ryan Holiday, "The Obstacle Is the Way."

Above all, stream of consciousness exercises help you get into the zone. Mihály Csíkszentmihályi calls this state **flow**. Flow is an optimal experience for us humans, where we work on a challenge, but we have all the necessary resources to meet this challenge. In a flow state, our perception of time is lost, because we are so deeply immersed in the activity. Based on several surveys, flow experiences count as our most enjoyable life experiences.

Recording a good video is a flow experience. If you are interested in reaching the sweet spot that enables you to record videos, check out my book *The Charismatic Coder*.[8]

If you prefer not recording a video introduction on yourself, an alternative solution is a relevant case study for your employer. You can write about a

---

[8] Zsolt Nagy, *The Charismatic Coder* (Leanpub, 2018). https://leanpub.com/The-Charismatic-Coder

challenge that could be relevant to your employer, for instance, establishing knowledge transfer in a team of X developers and increasing and measuring user experience of your web application to create more satisfied customers. Make it authentic; make both your hiring manager and your technical interviewer be interested in reading your attachment. You can do this by delivering a hook point in your cover letter, such as: "Teams work optimally if they are on the same page when it comes to sharing knowledge. Many teams struggle in this aspect. These are the steps we have employed to make knowledge transfer as effective as possible. I would love to compare our knowledge transfer approach with yours during the interview. Feel free to read about my case study in the attachment".

Submitting a case study sounds crazy on some level. It will not always work with all employers, especially if your employer has bad working conditions.

When giving your employer a free gift, you use the principle of **reciprocity**. In exchange for a free gift, your employer is not only encouraged to interview you, but they will also feel that they should talk about your topic, let alone the fact that you have qualified them with this approach, suggesting that you want to work with companies who want to encourage teamwork and knowledge sharing, without writing the cliché that you are a great team player.

Writing resumes is hard. This is why we started with your personal brand first. You will be able to showcase elements of your personal brand in your resume.

You will be able to create an interesting resume by knowing what your employer wants and offer satisfying their needs in an interesting and readable format. In order to make your application package interesting, you can apply a storytelling process. Shift focus from data and enumerations to stories.

Apply storytelling for your cover letter too. Talk about the expectations of your employer, and connect your cover letter with your resume. Make your audience interested in reading both your cover letter and your resume.

Always tailor your application package to the needs of each company you apply for. You will spend more time on preparing your application package per company, but you will save time in the job hunt process overall with a targeted approach. You will also get a chance to earn more than those who choose to bomb hundreds of companies with the same resume.

Consider doing things that no one else does. You may want to submit a video with your application package or create an attachment demonstrating a case study of a project or an interesting and relevant problem you have solved during your career.

Exercises

Read the next section "Social Media" before completing these exercises! You might want to consider creating or polishing your LinkedIn profile before creating your resume.

1. Review three job posts that match your profile. Take notes on what your employers are looking for.

2. Create three stories about three of your most important accomplishments from the perspective of your current specialization.

3. Update your resume inspired by this section.

4. Select the best position out of the three positions you created. Make an exhaustive research on the company. Then tailor your resume based on your research. Write a cover letter to complete your application package.

5. Record a short 2–3-minute introduction of yourself with your camera or phone. If you don't have relevant experience, practice it until you can talk 2–3 minutes in front of a camera without interruptions. This may be a hard task, it may take hours or days, but it's worth it from the perspective of your self-esteem and your career.

Social Media

Goals of this section:

- Focus on your LinkedIn profile for maximum results.

- Clean up your Facebook profile.

- Extra credit: Tweet or post about technology.

People respect those who share their thoughts. This is why you can enrich your personal brand by going the extra mile and creating a social media presence. You will be able to form meaningful connections, and you will also add social proof to your personal brand.

Ask yourself the question: What is the essence of your message? Check your vision and mission statements, and focus your efforts on this area. Remember a brand is a promise, and your audience has expectations. You deliver your promise by creating consistent messages.

The more platforms you use to share your thoughts, the better. This is why this section is about social media.

We will focus on the following profiles in this book:

- LinkedIn
- Facebook
- Twitter

The most important social media account from the perspective of your career is LinkedIn. In the 21st century, you will hardly get away without having a LinkedIn profile. Other profiles may also be important, but I would encourage you to prioritize LinkedIn above everything else.

LinkedIn: Your Online Resume

Most employers will find your LinkedIn profile. Make sure you reach all-star status.

Given that all principles of personal branding and writing a resume apply for creating your LinkedIn profile, this section builds on the principles of writing a resume. However, it is advised to create your LinkedIn profile before creating a resume. This is because your LinkedIn profile will contain a superset of information that fits in your resume. You will present yourself from the angle of your specialization, regardless of what a specific job ad may require.

Your LinkedIn profile is still targeted, and you still have a target audience; however, this audience is generalized to the extent that your LinkedIn profile alone is hardly enough in an application package.

In fact, I would not encourage you to export your LinkedIn profile into a resume, as it is a lot less effective than sending even a below-average resume with a link to your LinkedIn profile.

Make sure you base your resume on your LinkedIn profile, and keep them consistent. If there are major differences, it will mean a major disadvantage to you. Regardless of whether the inconsistency is a mistake or a lie, regardless of whether your employer asks you questions about it, the damage is already done by the time they discover the inconsistency.

One of the most important elements of your LinkedIn resume is social proof. If you form meaningful connections with other credible people and they recommend you, your chances get better.

Some hiring managers told me that they did regular consistency checks on LinkedIn and asked people who recommended a candidate to confirm their skills.

I can still recall the time when a CEO of a company cross-checked me while I was interviewing with them. Without him mentioning this to me, he got in touch with my manager from another company. Fortunately for me, I had had an excellent working relationship with my manager, and therefore he even contacted me and asked me if there was anything special worth mentioning that would increase my chances.

As you can see, a LinkedIn recommendation has to be credible enough so that it endures the occasional test of your potential employers inquiring about you.

Here are some LinkedIn tips:

- You have more space on LinkedIn than a traditional resume. The LinkedIn format makes your content easy to scan.

- Rewrite your whole profile whenever you change your specialization. Make your unrelated accomplishments less prominent, and make accomplishments linked to your current specialization stand out.

- Ask for recommendations for your past roles, whenever you have a good connection with some of your past colleagues. The recommendation of your manager or CEO is worth more than a recommendation of your peer.

- Order your skills from the perspective of your current specialization. You can override the default ordering sequence of LinkedIn to encourage votes on your skills. For instance, businesspeople tend to vote for PHP simply because they have heard about it. If you hide PHP among your skills, or you make it invisible, you will get the right upvotes.

- When your position changes, it makes sense to create a new entry in your employment history; otherwise, the description of your accomplishments will become blurry.

Facepalm or Facebook?

In theory, companies are not allowed to discriminate based on attributes of your private life. In practice, hiring managers often look at your Facebook profile. Even though many companies respect the boundaries between our professional and our personal life, there is a limit.

For this reason, make sure you don't expose photos, videos, and posts that act against your personal brand. If you have controversial content, make sure only your friends can see them. This does not make you a better or worse professional, but we want to eliminate the noise in the mind of our hiring manager. We want our hiring manager to focus on our professional life.

If your life is interesting and your lifestyle catches the eyes of other people, your Facebook profile becomes a benefit. It shows other people that you are an interesting, down-to-earth person. Remember one attribute of influence is if people **like** you. You are more **relatable** with your hobbies.

Facebook is hardly ever used for professional reasons, unless you create a Facebook page about a side-project or side-business. This section is not about using Facebook to advertise your technical skills. This is rather about making your personal profile a bit more relatable.

Twitter

Tweeting about technology on a regular basis, and retweeting tweets of influencers, shows that you are following trends. The side effect of this activity is that you may get a follower base that acts as **social proof** in your favor.

If you launch a side-project, always tweet about it. If you post anything on Hacker News, tweet it. If you read something interesting on Hacker News or on a news site, tweet it. If you attend a conference, tweet it. If you speak at a conference, prepare several tweets.

It is absolutely fine if you use Twitter for personal reasons too. Just make sure you don't ruin your reputation with controversial content. Remember interesting topics build your reputation. Controversial topics, such as sexual content, politics, destructive criticism, and hate speech, won't help you.

Follow the right influencers in your field, so that you get access to news faster. You may even use Twitter for networking purposes.

What About Other Social Media?

Feel free to use them for personal reasons as long as they don't destroy your reputation.

My final advice to you is that you should not waste a lot of time on social media from the perspective of your career. Hours of posting just does not pay off. Creating an all-star LinkedIn profile and getting some social proof is more than enough.

Exercises

1. Create or polish your all-star LinkedIn profile.

2. Check out your Facebook profile, and develop a strategy about what content should be public and what content should only be seen by your friends.

3. Locate the main influencers of your specialization, and follow them on Twitter. For the next month, build the habit of tweeting original content at least twice a week and retweeting interesting news at least twice a week.

Choose the Right Side-Projects to Build Your Personal Brand

Goals of this section:

- Find out what 90% of the developers do wrong with learning.
- Create a motivating, cutting-edge learning plan.
- Learn how to build a shiny portfolio.
- Build a blog and write about things you learned.
- Understand why strategies of many other developers are wrong for your goals.

Many software developers tend to pick up bad learning habits. These habits not only make them spend their time in the most inefficient way possible. They also become very insecure and frustrated during the process. You can become more effective than most of your colleagues if you respect the following two rules:

1. Don't learn everything! Learn only what you need.

2. Create meaningful side projects that build your portfolio.

Many developers stress themselves out by trying to learn everything. Recall that in Chapter 2, you learned that comparing yourself to others is an unhealthy strategy that leads to frustration. Trying to learn everything often comes out of the fear that someone else already knows a skill you don't know yet.

You will combat these fears by creating a learning plan that supports your short-term and long-term goals. You will apply the Pareto principle and focus on the 20% of tasks that deliver 80% of your results.

Choosing the right side projects is also vital when it comes to accelerating your professional growth. Building another framework makes little sense unless you have groundbreaking ideas. Oftentimes, people create frameworks out of the motivation of learning something new. The reward for this effort is learning experience that you could have collected faster by working on something meaningful.

The first step in finding the right side project for you is to create a training plan.

Your SMART Learning Plan

The goal of this section is not only to plan what is useful for you. We will also create leverage by linking your learning goals to your career plan.

When will you know that you are ready for the next step in your career? As soon as you create something tangible. What is a tangible result? Something that you are comfortable with presenting to the outside world. A high-quality blog post, an online service, and a GitHub repository are great examples for this.

Back in the university, I was a great learner, and others thought I was a smart student. According to my own judgment of my past performance, I strongly believe that I was a student with average skills and way above average persistence. My motivation came from my achiever coping strategy I had developed as a child.

Note Babies are vulnerable, and their lives depend on receiving love. This love is initially unconditional, because the baby cannot do much for their parents. Sooner or later though, parents form expectations, which make the kid feel not loved unconditionally anymore. A 2-year-old kid cannot tell the difference between the statements "My action is bad" and "I am bad." Therefore, when failing to meet parental expectations, the mind of the kid protects himself or herself from experiencing the thoughts "I am not enough" or "I am not worthy of love." This protection comes in the form of a coping strategy the kid chooses. This is how some of us become rebels, some of us withdraw as hermits, and some of us become pleasers. A popular way to please is to achieve.

Both the hermit and the rebel have a hard time with learning and cooperating with others. The former has a tendency to give up on cooperation, refuses to ask for help, and withdraws to his or her comfort zone. The rebel actively debates and disagrees with their mentors. Both coping strategies prevent people from cooperating with others.

The pleaser has a harder task though, because the real problem of a pleaser is **productive procrastination**. Productive procrastination is the act of appearing productive, but avoiding the necessary uncomfortable steps that lead to success. An achiever's goal can be to ace a test, read or watch a learning material from end to end, or get certified. A pleaser pleases the instructor, their colleagues, but rarely themselves. Activities of achievers and pleasers often have a tendency to avoid addressing the most important but uncomfortable steps and justify postponing these actions with achieving or pleasing someone.

Following Ryan Holiday's great book, *The Obstacle Is the Way,*[9] regardless of your coping strategy, bear in mind that the type of challenge that feels the most uncomfortable to you shows you the path that leads to your own success. Coping strategies trick us into believing that we can achieve, please, rebel, or retreat to succeed. In reality, we just need to take ownership of creating tangible things that others appreciate, and learning new skills becomes the side effect of doing useful things.

As an achiever, I learned what it takes to ace tests. I developed an extraordinary reputation at my university that even surprised me from time to time. For instance, I can recall an exam where I had to talk about one topic I randomly selected out of 20. This was near the end of my university years.

Before the exam, a year of acing every single subject was behind me. Instead of the mandatory 60 credit points, I completed more than 80 credit points' worth of subjects. I also won a scientific competition, I got hired for an EU-funded university research project, and I started writing publications on Semantic Web. As a result of my past accomplishments, I not only developed my reputation, but I also had a need to relax and start living a life. This was the time when I doubled down on becoming a better go-kart racer and my dating life also got spiced up. In other words, I started living a well-rounded life.

Unfortunately, my achiever tendencies continued haunting me both at work and in other areas of my life. Before the exam, I concluded that I had no time to learn this subject properly. I didn't attend the lectures, and I only studied for 2 days. Yet, I had a good command of most of the topics in the context of just talking about them. There were just two topics I was uncomfortable with.

Before I entered the examination room, I saw some of my peers get their results. I will use the grading system A for acing the topic; B, C, and D for getting the second, third, and fourth best result; and F for failure. My peers got mostly C and F results. One of my achiever friends spent a week on this subject and attended most of the lectures. Of course he got an A.

I went in the room. My professor greeted me, as he knew me well from another side project. He invited me to randomly pick a topic.

Most students have a story when they learned 19 out of 20 topics and they picked the 20th. This was my story of tough luck. Fortunately for me, I just read my outline before the exam, so I started jotting things down before I forgot them.

---

[9] Ryan Holiday, *The Obstacle Is the Way: The Timeless Art of Turning Trials into Triumph* (Portfolio, 2014).

Once I finished taking notes, my professor asked me to present the topic. Within a few minutes, he must have found out that I didn't know much about the topic. Out of his good will to help me, he asked me some random questions from other topics. He could detect traces of me learning the fundamentals here and there, but with my professor's experience, he also knew that I did not understand much about his subject.

As the exam progressed, it was more and more clear to me that I could fail this exam. This was an uncomfortable feeling, because I was not used to failing any university exams in the past. I noticed my professor got more and more uncomfortable during the exam as if something started bothering him deeply. He seemed afraid of something. He asked one last seemingly easy question, I could at least relate to it in some way and contribute a bit, but the experience was missing on my end to use the material in a constructive way.

After my evident failure, my professor expressed that this was a very uncomfortable moment for him. I was prepared for the bad news. Maybe a D, maybe an F, who knows, either way I knew I had to retake the exam. But then, to my surprise, he continued, "I could ask you more questions for the A, but I feel the best grade I can give you now is a B."

The moment was uncomfortable for me, because I couldn't achieve the A. What I knew or what I didn't know didn't really matter to me at that point. However, later I rationalized this B in my degree won't matter much. I don't have to retake the exam anymore. I am good enough.

As I am writing these lines, I can now clearly see that I was the villain of this story, not the hero. I was also the main loser, because I only tricked myself. In the group of my professor, some students got three times higher starting salaries after university than me. They deserved it, because they didn't focus on getting good grades. They focused on becoming good professionals at the expense of getting bad grades from subjects that didn't matter to them. Until this year, I got good grades from everything, at the expense of becoming exceptional in the areas that mattered.

I mastered delivering results in the form of grades, research papers, and scientific competition wins. I also mastered forgetting many things I had learned, because after the exam, I didn't continue with my spaced repetition to retain knowledge.

This specific subject did not matter to me much. However, years later, I realized that there were major gaps in my knowledge, because I focused too much on getting good grades, sometimes without creating anything useful. It took me another few years to close the gap by participating in interesting projects. Today, I don't even look at a training material to absorb knowledge from cover to cover. I look at training materials as a reference to help me accomplish my goals. You will now learn how you can do the same.

Your learning plan will consist of five steps. If you read the first letters of these steps, you get the word smart. From now on, we will refer to these steps as the SMART learning plan.

1. **Sync**: Sync your learning plan and your career plan.

2. **Milestones**: Create the high-level steps of reaching your goals.

3. **Analyze**: Break down these milestones into smaller actionable steps.

4. **Refine**: Get feedback and iteratively improve your project based on what you learned.

5. **Test**: Verify your knowledge by presenting your deliverable to your audience.

■ **Note** The SMART learning plan applies the principles of project-based learning. We learn just enough to create something tangible. This learning experience does not necessarily provide you with complete lexical knowledge. However, you will learn how to solve meaningful problems in practice, you will gain confidence by completing your projects, and by the end of this experience, you will also get tangible results that you can present in your portfolio.

Let's see the five points one by one.

Sync

In order to learn a skill, you need to know why you want it. Get clear on your goals first. Ask yourself the following questions:

1. What are your career goals?

2. What skills will you need in order to reach these goals?

3. What resources do you need in order to learn those skills?

Focus on both technical and non-technical aspects of your career. Focus on multiple timespans such as the following:

- What are your career goals in 5 years?

- What needs to happen in 3 years to meet your goals in 5 years?

- What needs to happen in 1 year to meet your goals in 3 years?

- What needs to happen in 6 months to meet your goals in 1 year?

- What needs to happen in 3 months to meet your goals in 6 months?

- What needs to happen in 1 month to meet your goals in 3 months?

You may need to learn a new programming language, leadership skills, presentation skills, software architecture, functional programming, acceptance testing with Selenium, or anything you know you will use in the near future.

Once you are ready with your list, answer the most important question of your learning plan. What is the first, most important step you should take? Make the next step clear to yourself, and start focusing on it.

Milestones

For each skill you would like to improve, determine projects and activities that measure your knowledge. Launch projects that will put your new skill into action. Start with the why, and the what will fall into its place.

Make these projects useful to you. Building an app for real users is better than building something just for yourself. Teaching others, blogging, shooting a YouTube video are all great ways of measuring your progress.

Make sure your goals and milestones are tangible. Don't come up with a goal like "I want to learn React by understanding the example in the book I purchased." Most likely, you would spend most of your time trying to learn nitpicky details that you would forget within a week.

Many books and courses are excellent. I have learned from many of them myself too. However, if you finish a book without creating something new that goes beyond the content of the book, your knowledge of the subject will fade as time passes. If you push your boundaries beyond the theory, you will learn some skills that may stay with you for the rest of your professional career.

A reasonable goal may be the following:

> Create a single-page application in React and ES6 that enables software developers to track their own productivity using the Pomodoro technique and other useful productivity hacks such as (...).

Feel free to refine the exact specification later.

Your projects also keep your work focused. You won't waste time on unimportant details. You will learn what matters. When I wrote *ES6 in Practice*,[10] I had three guidelines in mind:

- Each section tells you just enough to get started with solving exercises.

- The exercises tell you stories and challenge you with a continuously increased difficulty. You will not only feel good after solving some exercises, but you will also know that you can solve problems on your own. Some of the exercises include concepts other books would give you in a theoretical section. By solving such an exercise, you prove yourself that you are capable of learning new concepts once you need to use them. This should give you confidence in your abilities and deepen your understanding at the same time. Worst case, you read the solution and learn as much as from other tutorials.

- You can develop your interviewing skills by including tech interview exercises as well as homework assignments.

Be specific about the time you are willing to invest in your side projects. People tend to underestimate the time needed for a tech project, often by a factor of 2 or 3. If you figure out that your lifestyle does not support creating a complicated side project, you have some life decisions to make. You will either change your lifestyle or your career plan, or you will have to create a different learning plan for yourself.

Your goal may be quite simple. For instance, if you learn React and you already have a blog, you may want to set yourself the following targets:

- A. Create a boilerplate for a React application, and upload it on GitHub

  - Skills: ES6, React fundamentals, Webpack, NPM, Babel

  - Deliverable: React boilerplate on GitHub

- B. Use your React boilerplate to implement a Todo application on todomvc.com

  - Skills: More in-depth React knowledge

  - Deliverable: A refined React boilerplate based on the learnings of creating an application

---

[10] Zsolt Nagy, *ES6 in Practice* (Leanpub, 2018). Check out www.zsoltnagy.eu for free chapters of the book.

C. Write a blog post about your learnings

- • Skills: More structured React knowledge
- • Deliverable: Blog post

It may happen that you set a big goal for yourself and you deliver your solutions incrementally in multiple milestones. For instance, in order to realize a productivity app, you might want to create

A. An abstract productivity model to implement:

- • Skills: Pomodoro technique, Kanban, productivity research
- • Deliverable: Productivity model, documentation, software specification

B. A responsive web site with components:

- • Skills: Bootstrap or premade web components
- • Deliverables: The skeleton of the web site with static data

C. Automated testing and mocking, which are most likely needed for development:

- • Skills: Mocha, Chai, SinonJs, API endpoint design.
- • Deliverables: Automated testing framework connected to the application and a way to intercept and fake API calls. Design your API endpoints.

D. A React layer:

- • Skills: React, NPM, Webpack, BabelJs
- • Deliverables: None, as it makes perfect sense to do D and E together

E. You may want to manage the application state with a library like Redux:

- • Skills: Redux or the Context API of React
- • Deliverables: A fully working application with a fake server

F. You may need some server-side code and a database:

- • Skills: NodeJs, MongoDB
- • Deliverable: A fully working solution

As you can see, in these six milestones, you may be learning an avalanche of technologies. Set some target dates for yourself by estimating the amount of work needed for finishing the application.

Make sure you don't treat yourself too hard for missing your target dates. The importance of measuring time is to figure out how good you are with estimating your own development velocity.

A side project like this will keep you busy for a long time. However, you will reap the rewards once you can demo your application. If your application is useful enough, you may even be able to monetize it.

Preparing you to monetize your application is outside the scope of this book. In order to develop an application for making money, you will have to continuously test what your users want and aggressively modify your product. This process does not create an optimal learning experience for you, so let's continue focusing on your career for now.

Analyze

It is now time to break down each milestone. Create a table of contents for yourself about the subject just as if you were writing a book. Determine how to break down the new skill into small pieces.

Discover prerequisites and redundant topics, and make choices about what you will focus on. Don't worry about making everything perfect. You will have several chances to refine your plan.

Do your research from several sources. When it comes to technologies, you can visit the official documentation or check out the tables of contents of books, courses, YouTube channels, online tutorials, and blogs. You may even find helpful GitHub repositories.

Once you are done gathering information, make sure you organize your information. For instance, a table with the following columns will do:

- Module number
- Description
- Goals/tasks
- Resources
- Status

Do your best to detect dependencies between the different modules you create for yourself, and order your modules accordingly. Don't worry if you miss a dependency or two, as your study plan is lean enough to accept future iterations. Step 4 will be all about iterating your plan.

Link accomplishments to most of your modules; otherwise, you will have a hard time tracking your progress. The more you use your acquired knowledge in practice, the deeper your understanding will be. If you just read a book, expect to forget 90% of the material. If you take notes, you may forget 80%. If you write an article on it with example code, chances are you will only forget half of what you learned. As soon as you complete a project in practice, you are likely to remember the fundamentals for years.

Refine

Once you implement a module, you will face difficulties. Make a note of them, and modify your study plan bottom-up, starting with Step 3 and ending with Step 1.

For instance, you may figure out a dependency that blocks you. You may either reorder your modules or add new modules to unblock yourself. You may even figure out that some of your modules are useless from the perspective of the outcome.

Always be goal-oriented. Cross out modules that don't contribute to your goal. I can't emphasize the importance of the Pareto principle often enough: 20% of your effort contributes to 80% of your results. When it comes to learning, the right 20% will give you the solid foundation that you can build on later in your career. If you need to dig deeper, you will understand the topics you need in depth. Many professionals get stuck in the bottom 80% of things to learn, at the expense of doing the top 20% right. Digging deeper only makes sense when you know where to dig.

The most frequent modifications only affect the big picture of your milestone. This requires you to reiterate Step 3 only.

Sometimes you will discover a cross-milestone dependency, or you may replace technologies as you get to know more about the subject. For instance, you may initially decide on using a REST API for your project, but then you stumble upon GraphQL. Given that you already know about REST and GraphQL perfectly fits your project, chances are it makes sense for you to redesign the whole milestone. This way, you can make better use of your time and learn something new.

Some of your decisions and discoveries may lead to small changes in your career plan. Other changes may shape the description of your projects. Make sure you go back to steps 1 and 2 and update them.

Don't worry about changing your plan frequently. Flexibility is key in getting the best learning experience for yourself.

Your iterations may also depend on external feedback. Be open about your side projects. Show them to your colleagues and your friends. Post about them online, and discuss your discoveries in tech forums and communities. You will be amazed how quickly you can improve your skills.

Last, but not least, step back and look at your career path from time to time. Learning new things may unlock new possibilities for you. These are paths that you had not been aware of. Some path may revolutionize either your future projects or even your current project. If you unlock a new path, don't be afraid of modifying your study plan in a top-down way.

Test

Test your knowledge by creating tangible results and sharing them with an audience. Let it be a GitHub repository, a portfolio site, or a blog post, find a way to show your results to an audience.

There will always be someone who loves what you do. There will always be haters too. As the famous urban dictionary phrase puts it, "haters gonna hate." Whenever you receive criticism, ask yourself what the intention behind the criticism is. Always take things objectively, and do your best to improve your products. Be thankful for constructive criticism.

You are also likely to receive destructive criticism. Don't be affected by it. If you are, go back to Chapter 2 and combat your fears and limiting beliefs. Criticism makes you stronger. Above all, facing criticism properly shapes your character, refines your professional code, and gives you a small edge at the workplace.

Always draw conclusions after finishing a side project. Ask yourself what skills you have learned and how you will use these skills in the future.

Your Portfolio

Your portfolio will be built in a natural way as you complete some side projects. There are two important pillars of your portfolio:

- Your GitHub profile
- Your web site

Some software developers say that they don't want to bother building a portfolio. This is absolutely all right. Some career paths will be locked for you, but you still have a chance to be a respected and very well-paid professional, with an excellent work-life balance.

Other software developers say that they force themselves to commit code every single day on GitHub. I wonder what happens to them during their holidays. Software development has been an integral part of my life. However, on the day when I drive 12 hours to get from Hungary to Germany, unpack my stuff, and return the rented car, chances are I won't be in a state of mind where it is useful for me to write a single line of code. I could go to the code base and pretend that I was working on something by committing a useless line of code. However, the most important person who knows about this trick, myself, would eventually make sure that I would feel awkward after an act like this. This is not the proper way to build self-esteem. Showing off that you can commit code daily is not worth the effort. Focus on increasing career capital, not on keeping up winning streaks. Find a schedule that is good for you based on your own life decisions.

If you don't want to host your own portfolio, setting up a github.io page for your side projects is easy. Knowing Git and GitHub is essential for most software developers anyway. If you don't know how to use Git, this may be the first learning topic for you. There are excellent free books and tutorials on this topic.

Web design is another area that holds people back from creating a portfolio site. I would not worry about web design. My first portfolio site looked really bad. It was just a bunch of links to my portfolio elements, accompanied by descriptions. I still got a lucrative position with it.

Back then, I didn't use templates and wanted to code everything on my own. Be aware that you can download a lot of free templates for portfolio sites. Just search for Bootstrap or WordPress portfolio templates on the Web. Alternatively, you can also host your portfolio on github.io and find Jekyll-based portfolio sites that you can tweak.

Is a portfolio necessary for getting hired? Considering employment, the answer is no, but it definitely helps. It helps you convince your employers that you have used something they need. The number one fear of employers is that they hire someone who cannot code. A good portfolio immediately eliminates this fear.

Note As a freelancer, your portfolio is more important than as an employee, because your clients are interested in the results you create and they rarely have a system in place to train you to add value. As an employee, the less experience and education you have, the more important your portfolio items become. Why? Because your portfolio items verify that you are just as good as a developer with a degree and some employers may give you a chance.

If your resume is good enough, you may think that it makes little sense for you to create a portfolio, as you get more than enough interviews anyway. I still encourage you to create one. Why? Simply because it makes you a better professional. Your portfolio is the side effect of your professional experience. After completing 98% of your study plan, it would be counterproductive not to add another 2% to show the world what you are capable of.

If you get enough first interviews, but you often get rejected afterward, the right portfolio may increase your chances of getting hired. If your interviewer reads about some of your side projects and he becomes enthusiastic about it, your interview may be a piece of cake.

Your Blog

Sometimes your learning plan does not give you the option of developing software for your portfolio. For instance, it makes little sense to create code based on Chapter 2 of this book. This is when publishing a blog article makes perfect sense.

Even when you finish your side project, a blog post on your conclusions solidifies your learning experience. Beyond structuring your thoughts, a blog post also gives you some publicity.

Many software developers have blogs. Oftentimes, they write four to five articles, and they give up. When I review an application with a 2-year-old blog, I consider it as a very small bonus. Even an outdated blog indicates that the creator learned something. When I see a continuously updated blog, I know that the owner of the blog is most likely enthusiastic about his or her profession. This enthusiasm is very hard to fake.

■ **Note** Popular content creators have advertised the trendy idea of publishing content on a regular basis and going for the long shot. The long shot entails that you can predictably grow from zero to hundreds of thousands of views a month by producing content on a consistent basis. If your dream is to get hundreds of thousands of followers and to monetize your brand, go for it. If you just want to get hired or you want to get more clients, chances are spending time on creating a blog post or two videos a week does not pay off for you. You also have to determine if you are in a mass market or in a niche market. If you pick a good niche by specializing, producing more content beyond a threshold offers diminishing return. If you actively market yourself, for instance, by submitting application packages, you need posts that showcase your expertise. If you want others to find you without networking, a blog will help you attract opportunities.

You might have heard the advice that you should update your blog on a weekly basis. Similarly to the idea of committing every day, I would not go this far. Creating a quality blog post is hard work. You don't want to commit to dedicating hours of your life just to create content. While creating a blog post on a weekly basis gives you some benefits, the answer to successful blog building lies in the Pareto principle.

Some of your articles will be great; others will only be opened by a couple of people. If you commit to a weekly schedule, there will be times when you are simply not in the state of creating cutting-edge content.

The justification for a weekly schedule is that your brand is a promise that you deliver consistently. However, by delivering below-average quality content consistently, you are destined to create a brand for yourself that you don't want. The plot of this book is about creating win-win situations. If a tight schedule for creating content makes you happy, go for it; otherwise, think about pleasing the most important person in your life: yourself. If you can keep up with a more relaxed schedule, not only you win, but your readers win too, as they will have access to higher-quality material.

A schedule still makes sense on your end; therefore, you might want to consider writing some articles in advance and publishing them according to your publishing schedule. This will make it possible for you to show up and produce quality code even if you're sick, and even if you drive 14 hours on that day. WordPress can publish your articles automatically, while you are busy with living your life.

■ **Note** A blog may attract many opportunities that you are not aware of. This book is a result of an opportunity created by my publisher who came across my work. I have created books and online courses, mentored students in online bootcamps, and secured many other opportunities. My blogs also helped me reach software developers with a significant problem in their careers, and I helped them level up in the form of mentoring programs.

These opportunities help you take more responsibility in your career, and this path goes beyond employment. We all reach our limits as employees, and beyond a certain level, gaining career capital points toward the direction of becoming an entrepreneur. This step enables you to set yourself free from depending on an external system, because you are the one who creates the system.

Based on my experience, I can help you on how to become a book or course author, a coach, or a consultant. I have been coaching and consulting clients since 2016, and my first attempt to create an online course is dating back to 2015. This experience has enabled me to distill the knowledge you need to get started in shaping your personal brand in the direction of becoming self-sufficient as an author, coach, or consultant. This topic goes beyond the scope of this book. If you are interested in this path, check out devcareermastery.com.

How Will I Get Readers?

This question pops up quite often. When developers start blogs, they don't know if anyone ever will find them and read them. Some developers start a blog, and they only get a few visitors per week. Absence of visitors can be very frustrating, and it can also lead to loss of motivation.

Most people don't know that it is easier to get visitors than you think. Most software development topics have forums, weekly newsletters, and boards, where you are welcome to share your content. One news aggregator is Hacker News (`https://news.ycombinator.com/`). The main advantage of Hacker News is that if your topic gets viral, it will get really viral. The downside is that it is not targeted at all for a given niche.

You can be certain that there are news aggregators in your niche. In the unlikely case that you can't find any news aggregators, you just got a great side project idea from me: build one. Targeted audience is more likely to visit your site than a big pool of software developers.

You can also experiment with Reddit. Be warned though that some subreddits ban explicit advertising of content coming from the same source. Therefore, you have to provide value by sharing other questions and content. Furthermore, some Reddit members are only there to give you destructive criticism. Other people are there to upvote trolling comments. Remember: just do your thing, continuously improve the quality of your articles, and whatever you do, remember "haters gonna hate."

Depending on your taste, Quora and Stack Overflow may be a bit more forgiving. In both sources, you may contribute to questions of your audience with your answers.

Don't worry about looking bad in front of some people. Your target audience will sympathize with you even if they see that the trolls attack you on a regular basis. Constructive criticism will improve the quality of your content. This is all that matters.

Does It Make Sense to Launch a YouTube Channel?

According to the mobile company Ericsson, video will make up 70% of mobile traffic by 2021.[11] This does not mean that seven out of ten pieces of content will be video, as we are talking about data traffic, not time spent on the Internet. The increasing trend of video consumption is still worth noting.

---

[11] The actual numbers may vary. You can check the latest updates on `https://www.ericsson.com/en/mobility-report`

Does it make sense to launch a YouTube channel? My answer to this question is "it depends." There are two prerequisites for uploading quality videos.

First, you have to structure your thoughts well. If you have a hard time writing a good blog post, why would you start recording a video? You will still take notes for your video, and sometimes you even write a script for yourself. This task is equivalent to writing a blog post.

Second, you have to combat your fears. Talking in front of a camera unites most fears that people are conditioned to feel: fear of rejection, fear of failure, and fear of public speaking. Talking in front of a camera also makes you experience how harsh your inner critique is. You may forgive a mistake if you hear it from others, but you will always judge yourself harshly. If you feel that combating all these fears and your inner critic makes you a better professional, the challenge is worth for you. Otherwise, at this stage, creating videos would just shift your focus away from what matters for your career.

A YouTube channel is not necessary for getting a job. It is not even the wisest investment of your time. A YouTube channel is not the best investment if you create apps for an app store or create web applications or web site. I mainly encourage you to create YouTube content if you want to become a coach, consultant, mentor, or author. To get a job or freelance opportunities, you are better off just creating a few quality videos to enrich your profile, or not creating anything at all.

Do I Have to Deal with Hosting?

There are many alternatives to self-hosting. I also started with a blog on github.io by forking a Jekyll template. It was a lean effort on my end, and my blog was up and running in hours. As I saw my audience grow, I decided on self-hosting with WordPress, using a free theme. You can do the same.

Hosting on github.io gives you the advantage that you can create a portfolio as well as a blog. The disadvantage of github.io is that no one will market your content for you.

If you want to get traffic right away, a nice publishing platform is Medium.[12] Two of my colleagues have written one Medium article each, and both of them got a lot of viewers. As they wrote about a trending topic, Medium just shared their articles with many readers. Even without a publication history, your first article may generate thousands of impressions.

---

[12] www.medium.com

What Other Benefits Will I Get by Blogging?

You will be contacted with opportunities from time to time. These offers may include the following:

- Job offers.

- Consulting offers.

- Asking for guest articles or podcast appearances.

- Other forms of cooperation requests.

- The opportunity to create paid tutorials for third-party sites.

- Some people may even ask you if it is all right for them to translate your articles into a foreign language. For instance, some of my articles got translated into Japanese.[13]

Some Notes About Focus

Focus on tasks that matter from the perspective of your career. Many people keep doing things that bring them no results. Once they find out that the results they wanted are not there, they try doing the same thing even harder. Doing the same things and expecting different results is nothing else but insanity.

Focus and having clear goals matter from the perspective of your career. Pareto-based personal brand building gives you the most results. Some tools listed here may enhance your knowledge, but they are not meant to give you a significant career advantage.

Stack Overflow is an excellent tool for cooperating on solving problems. Most software developers use it on a daily basis. However, answering Stack Overflow questions just for the sake of building a personal brand hardly pays off. I encourage you to help other people by answering questions, but don't develop a hidden motivation that your interviewers will be blown away by the quality of your answers or any opportunities will chase you. Writing on Stack Overflow is an advantage compared to nothing, but you can use the exact same effort to write some medium articles on popular topics.

Blogging in the wrong way may also hinder you. Creating content more often than what's comfortable for you not only drains your energy, but it also takes time away from your professional development. You will also lose your visitors quickly, as they realize the lack of consistent quality. Remember less is more. Don't keep up with a schedule when you don't have anything interesting to say. Quality trumps quantity.

---

[13] Example: http://postd.cc/javascript-debugging-tips-and-tricks/

Social media is great if you want to drive traffic to your blog. Social media is not a tool to build your personal brand with continuous posts. The only exception may be LinkedIn, where you have a chance to become an influencer and reach a lot of professional contacts with your articles.

Useless projects are time wasters. For instance, not many people want to see your Todo application in action, as the TodoMVC site has done a good job implementing it at least 50 times. No one wants you to reinvent wheels that have been invented many times.

Certifications are great for you if you are targeting jobs, where they are expected. In practice, most jobs never require you to get certifications. A few startups even view certifications as a disadvantage, as certifications don't certify that you can solve meaningful problems in practice. Only focus on certifications if it's a requirement for you. Otherwise, building your portfolio pays off faster.

Competitive coding challenges are great for testing and fine-tuning your skills. Tough coding riddles are great for practicing coding interviews. Coding challenges are often useless when it comes to building your personal brand. A few exceptions apply, as the winners get some job opportunities and badges that look good in your resume. Another exception is that some tech giants such as Google also recruit by spying on your search history of getting deeply immersed in some coding challenges. Competition is very tough. In order to get to the top, you need to invest a lot of energy into solving coding puzzles, and some of these puzzles correlate less with your job than gaining more experience in doing your job better. Hopefully you already know where this energy should be invested if your goal is to build your personal brand.

Exercises

1. Create your SMART learning plan.
2. Launch a side project.
3. Collect your past accomplishments and create a portfolio site.
4. Create a blog.
5. Drive traffic to your blog and portfolio, and see what happens.

Networking in Meetups and Conferences

Goals of this section:

- Understand the role of meetups, workshops, and conferences in your career.

- Consider becoming a speaker yourself.

- Discover that rock star developers are often not better than you in terms of technical skills.

Many people think that software developers are introverts and lack communication skills. Meetups, workshops, and conferences give you the chance to break this stereotype. Start small, attend some meetups, talk to people, and watch how well you can extend your comfort zone.

In this section, we will break down how you can make the best out of meetups, workshops, and conferences. You will learn how to utilize your time the best as an attendee, as well as becoming a speaker yourself.

Attending Meetups, Workshops, and Conferences

Once you visit a meetup, you will figure out that most people have no clear goals in mind. They just attend a meetup for the purpose of consuming content and maybe asking some questions. Attending a meetup for learning new things about technology is not the best idea. First of all, there is hardly enough time to go in depth; and second, you can watch the presentations afterward on YouTube or Vimeo at your own pace. Third, at least half of the topics will be off-topic for you from the perspective of your study plan.

The primary benefit of meetups is that you can form meaningful connections. Once you attend a meetup, you can network with other software developers, find out what they are doing, and find a way to cooperate with them. Unlike conferences, meetups are local, which means to you that you will have an easy time meeting other participants later.

The secondary benefit of meetups is that you can discover some new ideas that can help your career. The most precious information always puts things into new perspective for you. If you can take a lucrative idea home that makes you more effective, your time in the meetup was worth it.

Attending a workshop makes sense for you if you are active. Raise your hand whenever you have a question. If you get stuck, ask for help. Make sure you make the best out of the experience. If you paid to be there, you should also get something in return.

Workshops are also great places for networking. Talk to people, connect with them on a deeper level by finding out more about their motivations, and think about mutually beneficial ways of cooperation. Then exchange your contact details.

Conferences take everything to the next level. You have to go to a specific city, and the organizers do their best to give you the best possible experience. Some conferences charge you a four-figure sum for a seat, and you have to enter a lottery to get an opportunity to purchase a seat. Therefore, I consider it a major waste of money to go to a conference just for learning and consuming content. Networking is essential for you. In order to network, you have to be able to give something. This is why it is so useful to create an online presence in advance. Even if you don't have an online presence, you can at least give your attention to the other person and contact them later, once you know how you can cooperate with them.

Start small, and learn networking skills on your own by attending meetups first. Once you go to a conference, you will know what to do.

Don't worry if you are not comfortable with networking. As you talk to people, you will find out that everyone appreciates some attention. Everyone has the same favorite topic: themselves. If you show interest in them, your connections will often be interested in keeping in touch with you.

Networking is the best way to tap into the hidden job market without using guerilla techniques. A lot of cooperations are formed in conferences. If you have attended conferences just for the sake of learning, open your eyes, and see the opportunities that are out there.

Even though most people are interested in themselves first, don't be afraid of becoming a giver. I highly recommend reading the book *Give and Take: A Revolutionary Approach to Success*,[14] where you can find out that not only the least successful but also the most successful people are givers. No takers or matchers can replicate the success of a true giver. The main difference between a successful giver and an unsuccessful giver is that successful givers recognize where their efforts are spent the most effectively. They recognize takers and either stay away from them or encourage them to become givers themselves. They also give value that matters to others and still work effectively on their own duties. Don't be afraid of giving even if it does not come back to you.

Another book worth mentioning on the act of giving is Professor Cialdini's *Influence: The Psychology of Persuasion*.[15] By acting as a giver, others are encouraged to *reciprocate*, by finding a way to give something back to you. Givers give by nature. Matchers give out of their interest of matching your favor. Takers may give either out of guilt or out of their selfish interest of keeping a beneficial relationship going. Think about the last time when you were invited for lunch. How did you feel? Did you want to reciprocate the invitation?

---

[14] Adam Grant, *Give and Take: A Revolutionary Approach to Success*, reprint edition (Penguin Books, 2014).
[15] Robert B. Cialdini, *Influence: The Psychology of Persuasion*, revised edition (Harper Business; 2006).

Become a Speaker

Similarly to starting your YouTube channel, you won't become a conference speaker overnight. Your chances get better as you gain experience by speaking at smaller events and as you build your personal brand.

From the perspective of the conference organizers, it is evident that they want to work with rock star developers to sell more expensive tickets. If they have no proof that you can speak in front of an audience, chances are that they will reject you unless you have a very good topic.

In order to even out chances, and in order to provide fresh insights of new participants, many organizers anonymize the abstracts of the applicants and select the best ones without taking your personal brand into consideration. This gives newcomers a great opportunity to secure a speaker slot. Be aware that some conference organizers require you to submit at least one video where you show that you are capable of talking in front of a camera.

If you have a hard time recording a 1-minute pitch of your talk, don't worry. This is natural. At first, it took me more than half an hour to record a 1-minute pitch. I felt awkward and clumsy. As I gained more experience in talking, I felt more and more comfortable.

Talking in front of a camera and publishing the results feels good afterward. You have a chance to drastically improve your communication skills by talking to a camera on a weekly basis. Imagine what these skills mean to your career in the long run. You may even talk like a rock star developer on stage.

Given that most conferences want big names, some rock star developers also get some speaker slots without going through a formal application process. If you want to become one, it is a path worth pursuing.

Starting small makes sense. Speak at a meetup. You can even organize a meetup. Then offer your help in a workshop. Eventually, you will get enough experience under your belt, and you will be able to comfortably attend a conference.

If you feel that you are ready for a conference presentation, don't hold yourself back! Some of my colleagues have secured speaker slots with zero experience in meetups and workshops and no online presence. All they had was an interesting topic.

Don't worry about your presentation skills. Most conferences give you a package of tips to prepare a quality presentation. Your colleagues may also listen to your presentation in a dry-run.

My final advice to you on presenting is to let go of the internal judge. Record yourself and record another person. Watch both videos. Notice how harsh you are with yourself when you make mistakes. Notice how forgiving you are when the other person makes the exact same mistakes. Realize that your audience will be equally forgiving with you. All you need to do is give yourself the permission to make mistakes. You will be human; you will be more relatable.

Exercises

1. Find a niche that you are interested in. Develop a learning plan about the field of your interest.

2. Find local meetups that are related to your chosen topic, and attend some.

3. Create a short presentation and speak at a local meetup.

4. Consider securing a speaker slot for a conference. Challenge yourself by applying as a speaker to as many conferences as possible.

Summary

This chapter describes the toolset you can use to establish your personal brand. Opposed to many personal branding resources, this chapter derives your personal brand from your values. The values you consciously select for yourself determine what you want your career to be about. This determines your mission statement, your vision statement, your tagline, and your elevator pitch.

Once you know where you are heading toward, you will be able to select employment, freelance, and other self-employed opportunities that are interesting to you. Once you select employment or freelance opportunities worth pursuing, instead of spamming them with yet another average application they receive regularly, you can also tailor your resume and cover letter to serve the needs of your prospective employer or client.

To increase your chances, you may choose to create a portfolio, maintain a set of valuable contributions on GitHub, create an all-star LinkedIn profile, clean up your social media, start blogging, and even create a few videos. You can increase your professional network further by attending meetups and conferences for the purpose of networking.

Set Yourself Up for a Promotion

Unlock New Opportunities Using Your Skills and Attitude

In this chapter, you will learn how to develop and use your skills to unlock opportunities that result in raises, promotions, or new opportunities.

Although it may be tempting to skip this chapter and focus solely on negotiation for a quick win, notice that this chapter is an integral part of your negotiation attempts. If you can choose between learning good strategies before playing a game and just trying your luck, chances are that you succeed more easily with the former approach.

This is why we will focus on influence building, assertiveness, communication skills, feedback loops, and many other tools that come handy regardless of whether you are targeting a raise or a promotion, you are planning to interview for a new position, or you are planning to go self-employed.

© Zsolt Nagy 2019
Z. Nagy, *Soft Skills to Advance Your Developer Career*,
https://doi.org/10.1007/978-1-4842-5092-1_5

Social Skills and Assertiveness

Goals of this section:

- Discover the key of building lasting relationships.

- Improve your communication to assert yourself by taking the opinion of others into consideration.

- Minimize conflicts by learning nonviolent communication.

Your communication has a direct influence in your career. Regardless of the career path you choose, you will communicate with people.

While some jobs require you to communicate all the time, others let you get immersed in technological challenges. For instance, if you are an expert, a designer, or a software engineer, you may work in front of your computer for a whole day, without talking to anyone. Not communicating enough has a major effect on your social skills. Therefore, even if you have a job that requires little to no communication, make sure you socialize enough during your free time.

Dale Carnegie's classic book, *How to Win Friends and Influence People,*[1] is a bestseller on Amazon.com in the categories Leadership, Job Hunting and Career Guides, and Interpersonal Relations at the time of writing this book. Advice in this book is regarded as rock solid when it comes to developing soft skills and interpersonal relationships. Let it be salary negotiations, day-to-day business, or personal relationships, the generic principles outlined in this book will apply to most situations.

Dale Carnegie's book outlines some generic communication rules of thumb that will help you in your career. This is not a complete list, but some of my own takeaways after reading the book a few times:

- Think about what other people want. Help other people and show genuine interest in them.

- Talk in terms of the other person's interests.

- Never blame other people. Be very cautious about criticizing others. Never criticize people openly.

- Respect the opinion of other people. Even if you think they are wrong, never attack them. Ask questions instead.

- Whenever you are wrong, admit it quickly.

- Make other people feel important. Use their names often.

- Avoid long arguments. There are only losers in debates.

- Smile as often as you can.

---

[1] Dale Carnegie, *How to Win Friends and Influence People* (Simon and Schuster, 1937).

I suggest reading the book with a critical lens. Unfortunately, the book does not highlight the importance of determining your intent while using the principles. Each principle can come from a good place or from a very wrong place.

For instance, smiling often is great, as long as the smile is genuine. Everyone loves to be around smiling people. However, once the smile is fake, people will notice your subcommunication. It takes too much effort to fake your subcommunication; therefore, smiling often is rather a principle that suggests the transformation of your emotional state than becoming fake.

Regarding using the names of your conversation partners, appreciate it up to a certain extent. However, if you have not used the name of your boss in a conversation, chances are that implanting his or her name five times in your next discussion may make your boss think that you have a specific outcome with your conversation. There is a thin line between winning friends and creeping people out; therefore, each technique should be used with common sense in mind.

The earlier you figure out that communication is not about you, the better. If you keep talking about what you want, not many people will care. Help others first. If you want something, tell others what they get out of it. Blaming or criticizing other people can be lethal, especially openly.

Humans are able to detect emotions of others. For this reason, we are able to show empathy and feel what the other person is feeling. This is a skill that can be sharpened. Once you know what other people are doing, you will notice what feelings you trigger in them. The process of emotional state transference is an exchange process, where emotions are exchanged between the communicating parties. Positive emotional exchange requires emotional wealth. People with low emotional wealth rather drain the energy of their conversation partner. In order to avoid being an energy vampire, make sure that Dale Carnegie's principles come from a genuine, good place, not from a place of manipulation. Lasting success leads through empathy and emotional intelligence, while manipulation may only result in temporary success.

Note Entering into an emotional exchange with people who have low emotional wealth may burn you out. Unfortunately, people with cluster B personality disorders such as narcissism, Machiavellianism, and psychopathy may climb the corporate ladder quite effectively. If their agenda is discovered, they find another ladder using lies, manipulation, and nasty tricks. These people may recruit you or other people to support their agenda through faking vulnerability, sending you mixed signals, and using Dale Carnegie's principles in a deceptive way, telling you exactly what you want to hear. This process burns most mentally healthy people out emotionally.

Remember to protect your emotional wealth from predatory behavior and energy vampires, who are fueled by consumption while faking a genuine emotional exchange. If your boss is a narcissist, plays games with you, or shows signs of psychopathic tendencies, chances are your best option is to find a different job. If you are the manager of such a person, you need rock-solid advice to protect yourself, your team, and your company from the agenda of your subordinate.

Your actions have an effect on other people. For instance, as soon as you tell someone that they are wrong, chances are you will make their mood worse than before. Similarly, when you are in a good mood, you smile and you are energetic, people interacting with you will be energized as well.

Many people have a hard time accepting that the path to above-average people skills is avoiding debates and confessing when we are wrong. Since conversations should not be about us, it is not worth convincing the other person that we are right.

If you have an outcome you expect from another person, you are less likely to reach it than in an outcome-independent frame of mind. Therefore, people pleasers suffer the most when trying to apply Dale Carnegie's principles.

Note The principles of *How to Win Friends and Influence People* may be suboptimal for people pleasers. This is because (1) people pleasing is not a genuine act, as there is a hidden intention behind it, and (2) some principles encourage even more people-pleasing behavior, making a people pleaser seek clues about how others perceive their new strategies.

How do you treat a salesperson approaching you on the street or cold-calling you? I can sense the artificial communication strategies of salespeople easily, as I have been exposed to many of them throughout my life. I am still tricked from time to time.

Not long ago, I went to a company event in the United Kingdom. On the last day, I had to leave a few hours early on my own, because I had a coaching session scheduled for that night and I wanted to get home on time to avoid disappointing my client. Everything seemed easy, because I could save those hours by taking a train to London and flying from there.

As I was crossing a bridge not far from the Newport train station, I noticed a local greeting me in a friendly way. He saw me limping a bit due to a previous accident and offered a helping hand. He made me talk by inquiring about me; then he started sharing some of his own stories with me. I thought his interest was genuine; after all, he approached me in a natural way. After a few minutes, he quickly flipped his script and revealed his agenda. He wanted money from me.

How would you feel in a situation like this? Would you still appreciate the kindness of this person?

Let me share another story with you. An ex-colleague of mine requested a meeting with me. I had nothing against it, because we were in a good relationship in the past. We meet once every few years anyway.

He wanted to show me his new office. He was starting a new venture with his business partners. He had a hobby of driving a new BMW around, but this time, the BMW was missing. I asked him, "You didn't wreck your car, did you?" He replied, "No, I will tell you about it later." Mysterious, but fine, after all we were about to enter his new office and he looked excited.

We spent the next half an hour discussing about his business plans, while he was also inquiring about my life and my situation. Something weird was in the air. He knew how straight to the point I was, yet, he not only made me talk a lot about myself, he also went from one superficial topic to another.

After I came back from the bathroom, he offered me coffee. While we were standing, I noticed he mirrored my stance. I changed my stance, and he mirrored it in a seemingly natural way quickly. This was when I knew for a fact that my ex-colleague was not really interested in me; he wanted something.

I directed the conversation toward exploring what he wanted. Before he could deliver the pitch, he told me, "I don't really believe I will succeed, knowing you, but I will try regardless." He mentally gave up on what he wanted before he got started; and from then on, his interest toward me, the mirroring techniques, and everything else just fell apart.

How would you feel in this situation?

This is exactly what a people pleaser would cause trying to please others with techniques to win friends and influence people. Real social skills that win friends and influence people come from a place of genuine interest, not fake interest. We win friends by maintaining a positive emotional exchange, which comes from a place of abundance from the perspective of emotional wealth. People pleasing comes from a place of scarcity, because we want something we think we cannot have.

The Fastest Way of Getting What You Want Is to Give It to Yourself

Following the work of Abraham Maslow and Tony Robbins, we know we have human needs. We all want security, variety, significance, connection, love, and many other things.

According to my experience as a coach, mentor, and practicing development manager, suffering at the workplace mostly comes from an unmet need of certainty or significance. When an employee has a burning desire to ask for a raise or a promotion or they are considering leaving their company, there is an underlying need that is not being met.

It is rarely about growth. If you want to learn new skills, you can do it after work, or you can plan your progress and execute the steps needed to get ahead. Frustration and anger comes from the place of scarcity. We tend to think we are not enough and we suffer. I have been to this place too, and it feels like we don't know what we can do to change our situation. It seems our manager does not care about us. We may think they exploit us on purpose by underpaying us for our services. From this place, any attempt to negotiate a higher salary may backfire on you very easily.

At the same time, those who have a lot of options and are generally happy can focus on contribution. They get raises and promotions even if they don't ask for it.

Even if you get what you want, chances are you won't meet your need. How long does the emotional spike last after getting a raise? Weeks? Days? Hours? When I was insecure and a lot less emotionally mature than now, I can recall getting a 25% raise. I signed a contract in the afternoon, and I was already planning how to target my next raise once I got home.

I got what I wanted, yet, my human need for significance was still unmet. The 25% raise gave me evidence that I was enough for that salary. I was worthy of earning 25% more than before. I still felt insignificant as soon as I thought about others who earned even more than me. I call this the *human need paradox* – oftentimes getting what is not meant to meet our needs.

■ **Note**   The human need paradox can be resolved by meeting our needs from within. If you need to feel significant, you can feel your self-worth without conditions that depend on other people or circumstances. Circumstances do not define who you are. Other people do not define your self-worth. Using your skills to create value for others and for yourself is a lot more healthy way to meet your need for significance and certainty. As long as you are alive, you can be certain that you are able to create value. Therefore, you have a reason to feel significant.

This is why I warn you about using communication techniques. When they come from a genuine place, they are brilliant. When they don't, it's like putting a Band-Aid on a deep wound that needs to be disinfected. Even though your symptoms may ease, the underlying root cause could just get worse below the Band-Aid.

■ **Note**   Communication techniques are not meant to make you non-needy. You may succeed hiding your neediness temporarily at an expense of suffering. The only lasting way of becoming non-needy is to meet your needs from within.

You can feel significant by giving. You can also feel certain that you are on the right track by growing and contributing.

This strategy is great on paper, but there is one special ingredient that is missing. There are four types of situations: win-win, win-lose, lose-win, and lose-lose.

By giving, we ensure the other party wins. This can be a lose-win or a win-win situation. The optimal situations to seek in life are win-win situations. This is the interest of everyone, because if you live in abundance, you can give more from what you have. An optimal communication strategy for creating win-win situations is assertive communication.

Assertive Communication

Regardless of how well we take the interest of other people into consideration, our job is to reach mutually beneficial conclusions.

People communicating submissively hardly consider their own interest. They tend to say their own opinion does not matter that much and submit to serve the interest of other people. A submissive person hardly ever negotiates and gives up easily.

Aggressive communicators only tend to care about their own interest, even at the expense of the interest of other people. Not many people like aggressive behavior. Aggression tends to trigger defense mechanisms in others, shutting down negotiations.

We want to find a balanced way of communication between two lines by pursuing our own interest while taking the interest of other people into consideration. This is the basic idea behind assertive communication. If you communicate assertively, you strive for win-win situations.

In order to communicate assertively, you have to believe that your message is important, and your interest has to be clear. This is a challenge in itself, as many people have no idea about what they want in life. This is why they turn to manipulative negotiation techniques.

Assertive communication implies a degree of *vulnerability*. We transmit our own point of view, and we are open to discussions about it, without trying to convince anyone that our point of view is superior in any way. Our goal is to reach a conclusion, which is mostly somewhere halfway between your ideal outcome and the ideal outcome of the other person. Once both parties realize mutual interest, the parties are in rapport, and a conclusion is reached.

Oftentimes, understanding feelings is more important than facts. We tend to simplify and distort each other's messages anyway. For this reason, it is very important to practice active listening. Instead of participating in discussions, many people are just waiting for their own turn. Our goal is to understand the other party and transmit that their message is heard.

■ **Note** The main cause behind not understanding other people is ignoring how they feel. An easy way to become more empathetic is to consider feelings of other people as facts.

"Andy says he feels offended because you did not answer him in 5 minutes" is a true statement. In mathematical logic, it is considered a reification statement. These statements are supposed to be assertions in our knowledge base.

Accepting the feelings of other people as facts does not mean you have to agree with them. It just means you have to consider them as facts while considering your actions.

In order to encourage discussions, use more pauses; and make sure that before pausing, you end your message with a downward inflection. This is the *burst and pause* technique. You tell your opinion in a burst; then you pause and let the message sink in. Alternatively, your conversation partner may have a question which you may answer. Artur Schnabel, a professional pianist and composer, once said, "The notes I handle no better than many pianists. But the pauses between the notes – ah, that is where the art resides."

As you are reading, you may wonder why the downward inflection is important. First of all, you sound more mature. Second, if you end your messages with an upward inflection, your conversation partner will never know when you are done.

While communicating, other people are busy with their internal narration. We can get to know elements of this narration by listening actively. We can figure out what other people think about our opinion. The next step is to figure out what the other person wants.

■ **Note** The structure of internal narrations is considered an advanced topic. Internet marketers, copywriters, professional closers, and salespeople are forced to learn reverse engineering the internal narration of people to do their job. Software developers don't have this privilege. Bear in mind that the more neurotic a person is, the less they hear what you say and the more they hear what they want to hear. For instance, a person with narcissistic personality disorder may either perceive every statement that conflicts with their idealized image as a personal attack, or they may straight up refuse to hear and understand the message if they don't take your opinion seriously.

In the case of raises, our contribution is the product we are selling. There is a market out there, namely, the skill set required by your company, in order to solve meaningful problems. Concentrate on how you can give your market the best possible product, and you will increase your value. This is the idea behind the *product/market fit*. Understand the needs of your company, understand the services you are capable of providing, and make sure your employers understand how you can serve your company.

It is also important to identify how you can make your managers shine in everyday situations. Always make sure you take their interest into consideration; otherwise, you will run into hidden obstacles during any negotiation process.

In some cases, your peers or your manager may be in a long-term conflict with you. Fighting for power is not uncommon in organizations. I tend to say by nature that these types of organizations are not worth working for. However, you may be in a situation when you cannot afford to be selective. In this situation, being assertive is often the way to go. Think about the interest of the other party. By being assertive, you may discover that there is a solution even if someone originally wanted to fire you.

There is a way to reach a win-win conclusion even in the case of a lasting conflict of interest. Oftentimes, your job is to extend the scope of the discussion and find something both of you agree with. For instance, both of you want your company to succeed. If you can agree in some ways to make the company more successful, you will start developing rapport with the other person; and eventually, you will be able to understand each other well enough to cooperate in resolving larger conflicts.

The side effect of this process is that other people will think about you as someone who is mature enough to cooperate with. This attitude will open doors for you.

Always go for win-win situations. There was a situation in my career, when common sense was on my side in a technical decision, but two of my peers didn't understand the benefits of my proposed solution. Given that my manager had no idea about technology in general, I was the only person who fully understood the impact of this decision.

Later, my manager confessed that she considered firing me, as I was relatively hard to handle, because I stuck to my opinion and I didn't seem to cooperate well enough. Had I utilized assertive communication, her struggle of firing me would not have been there at all. Instead of trying to consider the opinion of other people and understanding their motivation, I simply took it for granted that my advice works better.

I only found out in retrospect that the reason why my manager considered firing me was that a colleague of mine started complaining about me. Exercising some empathy, the motivation of my colleague was more or less understandable. I left zero room for him to express his opinion. His vision about his future was that he regarded me as a domineering person, and he thought it would be significantly harder for him to shine.

Mistakes like this taught me how important assertive communication is. We could have privately discussed both opinions, and I could have coached my colleague about the right approach by asking questions and actively listening to his perception of the problem. This way, he could have derived himself that his proposal would backfire in the long run. Our decision would have resulted in no conflicts.

Later, I managed to clear this conflict with my colleague, and I apologized for the way I acted. I also told him that I considered the interest of the company and I did not want to ruin his reputation by any means. Eventually, I made sure we established common grounds for cooperating in the future, and I opened the opportunity for feedback in case he felt bad about anything. I made him feel important, and I involved him in all discussions that were important. In exchange, I won an ally.

Being assertive is a very important tool to remove obstacles. You might think that you are doing an excellent job, yet some people have other interests. Finding out about these interests and making sure that you do not step on other people's toes is just as important as doing a great job.

Assertive communication includes the act of saying no. One of the best books I have ever read on assertive communication was the book *When I Say No, I Feel Guilty* by Manuel J. Smith.[2]

Dale Carnegie suggests avoiding lengthy conflicts whenever possible. However, assertiveness requires you to enter into conflict with others from time to time. These conflicts may go out of hand if you don't possess the skill of resolving conflicts effectively. When a conflict goes out of hand, an assertive approach may easily turn into aggression. Therefore, it pays off to learn some techniques about managing conflicts.

Nonviolent Communication

The skill of settling conflicts is very valuable, as long as you are not a person continuously triggering those conflicts for no real benefits. Oftentimes, a conflict can be avoided before it surfaces.

In defense of your opinion, some conflicts are important. However, many conflicts can be avoided with social intelligence.

---

[2] Manuel J. Smith, *When I Say No, I Feel Guilty: How to Cope, Using the Skills of Systematic Assertive Therapy,* reissue edition (Bantam, 2011).

Marshall Rosenberg's work on nonviolent communication[3] lets you express yourself in a way that minimizes conflicts. This process consists of the following parts:

1. Observations
2. Feelings
3. Needs
4. Requests

First, make some observations. You can observe anything that's bothering you, including a flaw in a process or a response you are not satisfied with. For instance, someone not responding to your requests can be an observation.

The second step is about expressing your feelings. In this example, the natural feeling you communicate is that you feel neglected.

Feelings trigger needs. In the third step, you should be able to express that your need is not to feel neglected.

Eventually, you conclude the process with a request. For instance, "Could you please spend a couple of minutes with me to figure out how we can cooperate more smoothly?"

> Let's assemble the whole message: "Sam, I am struggling, because I have realized that it often takes days until I receive a response to my questions. This makes me feel neglected. I would need a faster reply to be able to support you on the project. Therefore, could you please spend a couple of minutes with me to figure out how we can cooperate more smoothly?"

Compare this approach with saying: "You are always neglecting me! I want my answer now!" Decide which message would make you cooperate more.

While nonviolent communication has its use cases, sometimes it feels a bit artificial. Especially in the communication of your needs, sometimes they are implied by the context, while other times they may feel a bit artificial. The best formula that worked for me is when I connect my need with a benefit the other person can enjoy when meeting my need.

Let's conclude this section with another example. Imagine that you are working in a Mediterranean country where the outside temperature is 35 degrees Celsius, but the air conditioner is operating at 18 degrees Celsius. The air conditioner is blowing cold air right into your face. Other people are comfortable in the cold, but you are not. What do you do?

---

[3] www.nonviolentcommunication.com/aboutnvc/4partprocess.htm

I noticed that the air conditioner is blowing cold air in my face (observation), which makes me feel very uncomfortable, as I am afraid of catching cold (feelings). I would like to feel comfortable in the workplace (need); therefore, could we agree on a warmer temperature that still makes you feel comfortable?

You would be amazed how easy it is to lose the chance of a raise just because of a childish fight about opening windows, accusing people of talking too loud, or agreeing on the room temperature. Whenever a debate becomes aggressive, other people may start sabotaging your progress.

Exercises

1. Test yourself by expressing your opinion in an assertive way. First, start outside your workplace in a low-pressure environment to build momentum. Then start formulating opinions in situations when you can create a mutually beneficial outcome.

2. During a conversation, pay attention on the burst and pause technique. Listen to clues about when the other party wants to talk or ask you a question.

3. Think about mistakes you made in the past, when you gave your manager a hard time. Draw conclusions on how to eliminate these mistakes in the future.

4. What conflicts do you regularly have in your workplace? What can you do to resolve or prevent these conflicts?

Becoming Irreplaceable

Goals of this section:

- Define your professional standards.

- Differentiate between skills, knowledge, and action.

- Let go of fears of making time estimations and learn how to estimate your tasks more accurately.

- Learn how a professional negotiates.

- Learn how manipulation and bragging hinders your career while implicit influence helps your career.

We have laid down the foundations of professionalism in the previous chapters. Let it be communication, work ethics, a study plan, or the ability to say no, strong willpower characterizes professionals.

When amateurs appear to have lost motivation, professionals still reliably do their best. Instead of random actions, professionals form habits that serve them. The daily routine of a professional boosts their efficiency and maximizes their energy, and they almost always appear prepared to do any job.

Although professionals have very high standards, they are not perfectionists. Professionals exactly know when the quality of their work meets their standards.

Professionalism is one of your greatest assets. You will feel better about your work, and other people will also treat you more seriously.

Note Some notes about the term "becoming irreplaceable." Some corporate employees tend to cover themselves by guarding know-how and refusing to mentor others. This way, they become a SPOF (single point of failure), creating negotiation leverage for them. I do not encourage this behavior due to the drawbacks mentioned in the manipulation section. According to our definition, becoming irreplaceable means that you add so much value that it does not make sense to replace you. According to my subjective estimation, not many people are willing to choose this path; therefore, becoming irreplaceable in this sense is easier than you would think.

If people take you seriously, you will have a lot easier time at the negotiation table. You either get more support or you have an easier time finding another job.

Define Your Standards and Stick to Them

Why are professionals taken seriously? First of all, instead of finding excuses, professionals take responsibility for their work. Second, professionals have an internal code of quality standards. While amateurs do whatever other people ask them to, professionals act according to their own quality standards.

A professional tends to take calculated risks. While amateurs are afraid of getting fired, a professional is expected to give advice regardless of whether the advice is appreciated or not. In other words, the opinion of a professional does not depend on expectations of others.

Suppose that an amateur and a professional are asked to complete a project within 3 months and both of them know that they would need at least 9 months for a professional solution.

The amateur commits to trying his best. In the minds of their managers, trying his best means that responsibility is now on the development team and they committed to finishing the project in 3 months.

In his book *The Clean Coder: A Code of Conduct for Professional Programmers*,[4] Robert C. Martin argues that the best decisions are made through confrontation of adversarial roles. This means that as a professional, it is your duty to stay assertive and avoid submitting to tactics of irresponsible managers.

The Clean Coder is a great addition to the previously mentioned book titled *When I Say No, I Feel Guilty. The Clean Coder* puts the principles of assertiveness in the context of software development.

When unprofessional developers say, "OK, I'll try," they often want to be the heroes of the day. This is a dream that never comes true. Becoming a hero is a side effect of being professional. Unprofessional work will not be celebrated, even if you meet deadlines with low-quality code.

When it comes to self-confidence and self-esteem, a professional does not need any celebrations. Professionals tend to derive their self-worth from within.

If someone aspires to become a hero at the expense of delivering an unprofessional solution, chances are it is an act of overcompensation for the lack of self-esteem caused by lack of professionalism.

Your self-esteem grows as you act according to your values. Values are what you focus on. This is why delivering an unprofessional solution or saying things like "OK, I'll try" is very dangerous to your self-esteem.

Under some circumstances, a professional is ready to get fired. Under other circumstances, a professional resigns before getting fired.

You have to recognize where it makes sense to start negotiation. If your manager sends you out on an impossible mission and they do not respect your opinion, you have no other choice than finding a different job. Think about it. If you say no, there will always be an amateur volunteering for the job. If you say yes, eventually, your manager will think the reason of low-quality work is that you are not good enough. In such a situation, get ready to resign, and find a better job, where professionals are respected.

If you commit to being a professional for the rest of your career, you will have options by the time you feel like changing. In other words, you will be able to afford to disagree and leave tyrannical employers.

---

[4] Robert C. Martin, *The Clean Coder: A Code of Conduct for Professional Programmers* (Prentice Hall, 2011).

Do Not Get Defensive

In our culture, we often mix professionalism with knowing everything. My personal code has always been against putting too much emphasis on lexical knowledge. This is the reason why I tend to surprise many applicants in tech interviews.

For instance, once I assembled a recruitment funnel for hiring great software developers. The first step was a test. The surprising aspect of the test was that the score was not what I was mainly interested in. It was rather how the applicants completed the test.

I could see how much time applicants spent on a multiple-choice question. Three out of ten questions required the demonstration of problem-solving skills in practice. Most people got these questions wrong.

The remaining seven multiple-choice questions were there for the intention of slowing developers down. If they knew they spent 1 minute interpreting a question and selecting the right answer. If they knew how to Google the answer, they spent 1.5–2 minutes. If they didn't know what to look for, they could even spend 5 minutes.

Those who spent 40 minutes on ten multiple-choice questions only had 20 minutes remaining to complete the coding task. Those who spent 15 minutes on the multiple-choice questions had 45 minutes.

The coding task was not a standard one that could be found on Codility, HackerRank, or LeetCode. Therefore, Googling the solution was not too useful. Once again, I was checking problem-solving skills in practice.

I could also rewind how they got to the solution. Some people structured their code well, but failed to put the puzzle pieces together. They still advanced in the recruitment funnel. Others completed the task, but it took them a lot of time, and their code was a mess. I didn't invite them for the interview.

Interestingly enough, this step completely eliminated those developers who wanted to fake knowing something they are not skilled in. After this step, the success rate of the next interview skyrocketed. Since then, we never had to fire anyone during probation.

Focusing on utilization of skills instead of rehearsing facts pays off in an interview environment.

Professionalism comes with the knowledge of solving professional problems based on your experience and common sense. It is absolutely all right to refresh your knowledge in case you are not competent in a given field. People will get used to the weight of your opinion and know that sometimes you have to do your research in order to formulate a dependable opinion.

Even though lexical knowledge does not make you a professional, it still brings you forward. Most professionals share the value of curiosity. A professional tends to look everything up to fill their knowledge gaps. A professional spends hours every week studying. Building lexical knowledge is a side effect of professionalism.

Our mind works in a way that the more we know about a subject, the more associations we can formulate. The more associations we use, the better our problem-solving skills will become. People with good problem-solving skills tend to have above-average lexical knowledge.

Extending your lexical knowledge without putting it into practice is close to useless. This is why I encourage you to put your ideas into action and learn by doing. More associations will be created in your brain.

The Psychology of Time Estimations

Most people fear of estimating time needed for completing a task. One reason of this fear is *status anxiety*, dependence on the opinion of others. Our super-ego may tell us giving time estimations is dangerous.

In our fears, once we miss a deadline, we are regarded as less reliable. Not meeting deadlines appears to be highly unprofessional on the surface. Our inner critic will start blaming us for being late. Eventually, you will literally feel physical pain while announcing a deadline.

You have a chance to face this fear and shift your mindset about deadlines.

Let's explore the myth of deadlines. First of all, the problem with them is that the word deadline contains dead. This is a word with negative energy, and it also sounds dead serious. I hardly ever use this word, even when coordinating milestones. However, there is not much we can do about the word; it's part of the English language.

I normally propose using the phrase *target date*. It does not only sound soft, but it also emphasizes that it is only a target. In other words, in case something unexpected happens, the target date can be moved.

I do use the word deadline to highlight critical tasks. For instance, if a client asks us to complete a task in 2 weeks and they may leave us causing my company to lose 100 times my yearly salary, it is worth making sure that we communicate the severity of the situation. As long as you use the word deadline sparingly, other people will do their best, and they will also back you up.

When it comes to estimations, a professional creates best-case, average-case, and worst-case estimations. Depending on how the business reacts, out communication should be shifted accordingly. For instance, if they only listen to the best-case scenario, it is your duty to inform them that a best-case scenario is not a binding estimation. If they ask you to try your best, make it clear that "trying our best" is our default behavior.

Always check all the dependencies; plan your tasks properly. Apply tolerance on target dates. When something can go wrong, chances are it will go wrong.

Calculating with tolerance is necessary for handling unexpected situations. Professionalism comes with a good sense of implanting a tolerance factor in time estimations. This tolerance is not up to negotiation. If you say you "try your best" and you eliminate this tolerance factor, you are risking the delivery of your project in case anything unexpected comes up. This attitude is not cooperative and is highly unprofessional.

If your employer says your estimation needs to be modified such that you finish earlier, your best option is to start negotiating the scope of the deliverables.

Reducing the scope also appears in *The Clean Coder*. Robert C. Martin emphasizes the importance for striving for the best possible outcome. When an impossible demand is formed, a subset of the features can be deployed on time, within budget, meeting your quality standards.

If rescoping still does not lead to an agreement, your last resort is to start building technical debt while negotiating the repayment of this debt. Technical debt is a lower-quality solution in terms of maintainability. In blunt terms, this is what software developers call hacking skills. Make sure you manage the complete lifecycle of the software in the negotiation though. When you decide on building technical debt, make sure you also negotiate its repayment before taking it. Otherwise, business will consider building technical debt as a shortcut and may easily force you into compromising your options until your application goes bankrupt in terms of maintainability.

Using the terminology of the previous section, striving for the best possible outcome is an assertive behavior. The strength of your message comes from high self-esteem. You know that your message is important, and you are not willing to get into conflict with your own values.

If your employer forces deadlines on you that don't make sense, make sure you work on yourself in order to be in a position to resign as soon as possible.

Unhealthy time pressure compromises the quality of your work and the quality of your life. This may lead to the need for therapy, coaching, addictions, or recreational activities to compromise for this pressure, which may negate any salary benefits you get from a toxic environment. Don't end up spending money on artificially repairing the damages of your unhealthy employment situation.

Professionalism and Negotiation

Your actions should communicate professionalism and not your statements about yourself. If you wear a T-shirt with the message "Nothing to Prove," you are trying to prove that you have nothing to prove. This is a contradiction. Using the phrase "I am a professional" communicates that given you are not sure whether your employer knows you are a professional, chances are you aren't a real professional. Therefore, your standards should communicate whether you are a professional or not.

In other words, seek cooperation, not competition. If you show your employer why it is beneficial for them to apply your standards, they will be a lot more open and grateful.

Choosing the Right Tasks to Work On

Suppose that you can select between easy and hard tasks. Your manager has no idea about the difficulty level of these tasks, and he only sees the number of tasks you completed. Would you prefer choosing the easy tasks or the hard tasks? Why?

Whenever people are observed, they often want to shine. As a consequence, many people avoid hard and long tasks in favor of shorter ones, as they figure out that the path to their own success is through the KPI (Key Performance Indicator) of number of tasks solved per unit time.

In most cases, this is indeed a KPI, but only for not getting fired. In order to get promoted, you have to make a difference in another way.

Many people turn to blogs and training materials like this one, searching for the Holy Grail, wanting to move up the corporate ladder using frameworks and rhetoric. Internet marketers have figured out this trend too, and they are using interesting metaphors to sell products to you. Unfortunately, more people raise their hands to buy the magic pill that gives them a boost in productivity, performance, earning ability, and so on. Buying the magic pill is nothing else than spending money on a mere dream associated with effortless success.

Hopefully by now, you may have concluded that magic pills do not exist, because success is not meant to be effortless and instant. Your accomplishments do not make you happy. What makes you very happy or very sad is who you become in the process. The journey is an integral part of success, and therefore, any magic shortcut is just a way to trick and deceive yourself.

Internet marketers tend to sell you easy solutions to problems, where easy solutions are not meant to exist. The real magic pill most people need is taking responsibility. This is an uncomfortable truth; therefore, this advice only works in a niche market.

This advice applies to the question of whether you should choose the easy tasks or the hard tasks. There is no right or wrong answer. Obviously, meeting the KPIs you are hired for is important. Beyond that, the harder tasks may increase your skills and future earning ability.

In the short run, choosing tactics to meet your KPIs may make a difference in how others perceive you. In the long run, you reap a lot more rewards by going the hard way and tackling challenges that no one else dares to touch. Others may temporarily get into a good position by using tricks, Machiavellianism, and manipulation; but in the long run, these techniques are self-destructive. Therefore, avoiding manipulation is a wise career choice.

Defending Yourself Against Manipulation

Manipulation is a conscious effort in shaping our image in other peoples' minds for the purpose of them helping us reach our goals. Manipulation techniques are often superficial, and they demand extra value in exchange for nothing in return.

Note   People with narcissistic personality disorder focus on how they are perceived instead of doing good things. This is why a narcissist is dangerous at the workplace. They appear useful on a superficial level, but in reality, they use the unconscious help of their colleagues to fulfill their agenda. People recruited and manipulated by a narcissist are called *proxies*. A proxy is unlikely to know that he or she is supporting the secret agenda of a narcissist; they just act according to their own values or standards.

A proxy employee works like a proxy server. We may think we are interacting with the proxy, while in reality, we are interacting with the narcissist.

While it does not pay off to be manipulative yourself, it does pay off to be able to defend yourself from manipulative people. Recognizing manipulation attempts of others is important.

Note   Avoid emotional exchange with manipulative people, because the nature of the exchange tends to drain your energy. Manipulation comes from a place of low emotional wealth. When someone with high emotional wealth exchanges emotions with someone with a low level of emotional wealth, the result is mostly beneficial to the manipulator at the expense of the other party.

The interest of the manipulator is to apply a double standard and keep things vague. The manipulator holds you accountable for a strict standard even if you are their line manager. At the same time, the manipulator applies a different standard on himself or herself.

Note The easiest way to defend yourself against manipulation is by documenting all decisions in writing and leaving no room for creative interpretations. Similarly to software, facts and well-defined acceptance criteria do not leave room for undesirable interpretations.

Some social systems are less forgiving for manipulation than others. For instance, comparing the United States to Germany, US employment laws give more rights to the employer, while in Germany, employees tend to have a lot more rights. In some countries, firing an employee with an indefinite contract is illegal. In some countries, members of the works council are protected, and companies beyond a given size have to have a works council by law. In some environments, it is possible to fortify one's position to the extent that he or she can get away with predatory activities hijacking the corporate system for personal gains. This phenomenon can happen both in startups and in corporations, although it is less frequent in startups. When you encounter a person who appears to have hijacked the system, make sure you protect yourself.

Once I had a client who was asked to spy on other people by a person in the executive team. The exec indirectly suggested that the employee had no other choice if he wanted to stay in the company. Fortunately for him, he is at a different company now in a better position.

After learning how to protect yourself from manipulative entities, you may ask the question if it pays off to join the dark side yourself. Although this question would open a Pandora's box of philosophical debates on vaguely defined terms of art, I can summarize my generic opinion in one word: no.

The problem with joining the dark side is that you become vulnerable to distortions in your personality that you might not be prepared for. Manipulation comes from a place of scarcity. Scarcity comes from a place of neediness. Neediness comes from having unmet needs. Individuals with unmet needs do not know how to meet these needs from within. This leads to lower self-esteem and suffering.

You may feel upset, angry, frustrated, or bitter about other people trying to manipulate you or secure resources for themselves at your expense. I felt the same way until I realized that these people are just suffering, because they haven't found more empowering ways to meet their needs. There is no such thing as a good or a bad person, because these terms are relative.

After I started exploring the fundamentals behind IFS (Internal Family Systems) therapy, I realized that there is a frame of mind that helped me perceive intentions that had seemed negative to me in the past. The frame of mind I am using now is that every action has positive intent. The positive intent behind a manipulative action of a person is often protection due to a threat that feels real to him or her. Most people go through traumatic events, and trauma is stored in our body. If you are interested in this topic, the best book I can recommend is *The Body Keeps the Score: Brain, Mind, and Body in the Healing of Trauma*.[5]

If you want to avoid distorting your perception, your best defense is to take responsibility for your situation, do what is right, and avoid manipulating others to get what you want. This doesn't mean that you should not learn how the system works, because playing the game well to increase your utility benefits not only you but also the system. Being clever does not distort your perception. Hijacking the system for personal gains does.

Another problem with manipulation is that a company worth working for comes with intelligent and experienced people.

Note If someone is worthy of you impressing them, chances are they can see through your attempts of trying to impress them. This is why explicit influence tends to fail. You either impress the wrong targets, or you try to impress someone who can see through each and every attempt.

Have you ever worked with someone who keeps on worrying about his or her own image? The easiest way to recognize these people is their lack of ability to listen. While you talk, instead of listening to you, they rather listen to their own mindtalk: "Can't you see that I am the best? Isn't it obvious to you that I have to be rewarded for this act? Can't you see that I solved a hard problem again? Can't you see that I am mentoring my colleagues? Shall I repeat these questions five more times each day so that you finally reward me?" If you worked with a person like this for a while, would you promote him, or would you rather fire him?

By applying explicit influence, you run the risk of losing trust of other people, and you run the risk of undermining your own self-esteem. Your line managers may start questioning your integrity.

---

[5] Bessel van der Kolk, *The Body Keeps the Score: Brain, Mind, and Body in the Healing of Trauma*, reprint edition (Penguin, 2015).

Implicit Influence

Most people know when a manipulator is around, and manipulators rarely last. Let's focus on positive role models instead. Have you ever had very popular colleagues, who were not only professional but also said the right things at the right time?

These colleagues of yours unconsciously exercised implicit influence. Implicit influence is when you influence other people with your character, by adding value and having a trustable opinion. The reason why implicit influencers are trusted is that they are not even trying to influence you. Influence is a mere side effect of their actions.

Asking for a higher salary without adding value may burn bridges behind you. Going after your professional goals while respecting your values and standards is a quest other people will often back up.

Implicit influence is not something you can learn overnight. It is still possible to work on the factors that increase your professional charm.

Note Another term for implicit influence is narrative influence. This is because narrative influence causes others to develop an internal narration about you. Narrative influence lets your actions speak for you instead of you bragging about yourself.

The following checklist will help you work on your implicit influence:

1. Accept yourself in the way you are. Exclude manipulative behavior, and exclude all rules in your mind about what you have to do in order to appear better than you already are.

2. Practice formulating your opinion. Be able to express your opinion such that other people listen to you. Your opinion should not depend on what other people would like to hear, but rather on your best judgment. Always apply all principles of effective communication, and avoid arguments triggering negative feelings.

3. Improve your communication regularly by learning from your mistakes. Develop empathy. Watch what other people are feeling when you talk to them.

4. Gain expertise. Your opinion should not just be well formulated, but it should also be meaningful.

5. Be interested in solving problems professionally. Problem-solving skills are very valuable, regardless of the problem domain.

In order to be visible, you have to decide whether you go down the path of explicit or implicit influence. There is a natural paradox in showing off your skills to staying in the background.

Visibility is a result of the utility of your actions. If you work on important things, other people will want to cooperate with you, resulting in a positive internal narration.

Exercises

1. Write down your daily routine on a business day.

2. Make small gradual changes to your daily routine to define a schedule that enables you to perform professionally. What changes boost your energy, your attention?

3. Define your quality standards.

4. What would you do if you had to lower your quality standards due to external pressure? How do you communicate in these situations?

5. Think about situations when you felt you weren't acting as if you were a professional. How did it make you feel? What fears resulted in your compromise?

6. Think about three situations, when your own quality standards would make you disagree with the standards of your employer.

7. How does a deadline or target date estimation make you feel? If you discover any adverse emotions caused by a limiting belief, use the ABCDE method to eliminate them.

8. What learning goals do you have for the next 12 months? What skills would make you a better professional?

9. What are your main strengths at the workplace? What actions do you take on a regular basis to use these strengths? How do others react to your actions?

Giving and Receiving Feedback

Goals of this section:

*   Find out why feedback is important.

*   Make it easier for yourself to deal with negative feedback.

*   Learn how to give and receive cutting-edge feedback.

- Consider undertaking extra missions to accelerate your progress.

- Learn how to deal with constructive and destructive criticism.

On one hand, you have to deliver excellent work. On the other hand, bragging about what you do leads to explicit influence. Furthermore, doing things that are hard to notice and staying far too humble about your achievements will inevitably result in an average perceived performance.

Many corporate environments tend to apply a performance rating system. As it is hard to keep track of employee performance on a large scale, a performance rating simplifies this task by assigning a label to each employee. In some systems, there is a fixed ratio of overperforming and underperforming labels to be distributed. This leads to the unholy quest of distributing some underperforming labels to people who don't take it well at all.

Some tech giants tend to create competition by placing the bottom 10% of their employees on performance review automatically. Other companies just give the label "developing" to a fraction of their employees, which makes them feel undervalued. The corporation applying this standard does not mind the damage caused by such a label, because one of the following two things can happen: the employee either makes an effort to add more value, or the employee quits.

Unfortunately, in real life, things are a bit more complex, because life tends to get in the way of peak performance. We all have ups and downs in life. The loss of our loved ones, a breakup, or a health condition may have a temporary negative effect on our performance. Sometimes our manager does not even know the causes.

There is nothing worse than delivering a bad rating to a software developer who is highly skilled, but is not utilizing their skills to the extent they should. Once I tried to find some rationalizations to protect this developer with objective reasons that can be defended in front of a board that disputes each reason to slot a person in a category. This was when I got stuck, because I had little visibility of what this developer was doing. The minimum necessary feedback document was missing, because the developer didn't justify their own performance.

I collected feedback from some colleagues of this developer to make sure I am not making a major mistake with the rating. Then I started applying my constraint logic programming experience to assign the dreaded labels to some developers.

It turned out that those who built feedback loops and made sure others knew what they were doing performed better than those who didn't care about their self-evaluation document. In each and every case, their peers gave me mixed reviews. A pattern emerged as a common constructive criticism: lack of transparency.

■ **Note** Integrity is all about consistently executing your tasks according to your professional code at work. How you do one thing is how you do everything. If you slack off at your self-evaluation, chances are you are also slacking off in other areas of your work.

Let your actions trigger positive thoughts in other peoples' minds in a natural way. This is implicit influence. Walk the extra mile, think about your tasks based on common sense, and always think about becoming more productive.

Talking about being a professional, having high standards, and being efficient is explicit; and it is also artificial. Showing those high standards, tackling more tasks than anyone else, and boosting your own efficiency with tricks is implicit.

Implicit influence done right results in more trust and better cooperation. This is all you want to achieve in this section. Make sure your actions are indeed implicit and not manipulative.

You know that you got there as soon as other people start talking about you as their role model for doing something well. People tend to praise you naturally as you grow and make a difference.

Do the Right Things at the Right Time

Suppose that there is an emergency situation in your company, but instead of solving the emergency, you are thinking about a brilliant idea that makes your team more productive.

Even though you noticed the emergency first, you let the junior staff take over handling the situation. This results in 2 hours of extra downtime.

A day later, you present your brilliant idea to your lead, while he is still busy communicating about the emergency situation. Your idea is indeed worth saving thousands of euros of wages per month on repetitive tasks. Yet, the company just lost €20,000 due to the knock-on effects of the downtime. How would your efforts be perceived?

This example is artificial. In real life, I have heard of a person who signaled emergency situations quite well. As soon as an emergency was about to surface, he went away to take a long cigarette break.

Oftentimes, employees are bad at judging what is important to the business at a given point in time. When they selfishly take control of overriding what's important, disasters may happen. If you find out about an emergency situation, at least you should ask your lead about priorities.

Note One way to take more responsibility at work is to become more business-savvy. It is a lot harder to create tangible value to a business than to merely focus on technology. Many professionals refuse to accept this form of responsibility and escape in the maze of productive procrastination, delivering brilliant solutions that cannot be monetized, because the business needs something else at that point. By rejecting business-savviness, your earning potential and utility also becomes capped.

Jason, one of my clients, had limiting beliefs about money. He couldn't stand hearing about business-savviness. He believed that it was the job of businesspeople to find out how to utilize his efforts to make money.

Jason approached me because he wanted to transition from his software developer role toward software architecture, and he was looking for options to grow. I asked him if he liked the environment he was working at, and his answer was a clear yes. Jason loved the people he was working with, and he was sad about considering leaving them.

Then I asked him to identify some business needs where he could add value. This was where we got stuck. Jason refused to consider anything about the business. He was delivering software solutions, and he didn't want to hear about money.

As a natural consequence of his beliefs, I asked him if he is comfortable with earning more money. He said he was happy to accept more money if his manager valued his contribution. However, he would never ask for more money explicitly, because he feels uncomfortable with it.

Jason clearly had a limiting belief about money. On a superficial level, I could help him by asking him to find a mentor in the company who connected his skills with the business. Unfortunately, he was too shy about communicating what he wanted.

As a result, he had two options: he either finds a new job with more responsibilities, or he goes through an emotional transformation of letting go of his limiting beliefs about the business aspect of software development.

> **Note** A healthy employment relationship is a symbiotic relationship. The employee is able to earn more money as a member of an organization than alone. It also pays off for the employer to finance the employee's salary, because the work of the employee translates to a positive financial return either in the short run or in the long run. If any parts of the symbiotic relationship are not there, the employment relationship is not sustainable.

Some people have an excellent sense of what's important; others learn it by experience. Experience is gained via feedback. Your team lead or your manager is your client. Ultimately, the company is the client of your team. You are supposed to function efficiently, to fulfill the needs and wants of your clients. The only way to discover these needs and wants is through establishing regular feedback.

In order to be visible, take feedback into consideration. Doing the right things at the right time efficiently is one of the most important things you can do to get ahead in your career.

Feedback as Negotiation Leverage

Negotiating with implicit influence is hard, and it is sometimes even impossible. Yet, we can always build on the benefits of implicit influence and feedback loops.

At the negotiation table, instead of bragging about yourself, use your already established feedback loops. When you ask for feedback, you hear a polished manifestation of the internal narration of your manager. Regular feedback loops ensure that you know what actions result in honestly positive reaction. Instead of telling your manager how awesome you are, you can just ask for feedback.

In the case of regularly established feedback, it helps if you record some traces of what was said and when. Asking for improvement suggestions, and tackling them one by one, is a powerful tool. If you use feedback to your advantage, you will have a proven track record of continuously improving. This track record is worth a thousand times more than any form of explicit influence.

As a final remark, minimal explicit influence is often needed during negotiation; otherwise, how would you be able to assertively talk about your own interest? It may pay off to use explicit influence if there is already a consensus about your performance. This means that your statements about yourself are already in the minds of your managers.

Feedback is going to be the number one weapon in your utility belt for negotiating a higher salary. Feedback is not only essential in becoming a better professional, but it is also a great tool to find out what you need to do to make a difference in your current role.

You will become a more valuable member of your team by moving the whole team forward by exercising the fundamentals of giving professional feedback. You will also benefit from getting better at dealing with people.

The value of information you get through open and continuous feedback will also ease the negotiation process.

Encourage an Open Feedback Culture

Healthy companies have an open feedback culture. Ultimately, successful founders realize that the success of the company directly depends on their people.

It may happen that you are in a company where open and frequent feedback does not seem to be available. The consequences of no feedback are quite severe.

Even when everything is all right, most people develop their own internal narration. People may think their work is not valued or not appreciated or they are doing badly. Some people may be afraid of getting fired for no reason.

Fear may result in putting on some masks and hiding some aspects of your intentions and actions. These masks prevent smooth cooperation. Your mask hides parts of you that you know about, but others don't.

Fear also makes people blind in a sense of not being able to interpret feedback. This is the opposite of putting on a mask. Others see things about you that you don't.

The worst category consists of some aspects of your actions and behavior that neither you nor your surroundings know about. This unknown part can surface in debates and triggered mental states.

Self-awareness and an open feedback culture can decrease all three categories. Exchange of feedback will not only make you aware of some mistakes you regularly make but it also helps others find out more about you.

A famous model summarizes the above levels of perception: the *Johari window*.[6]

	Known to self	Not known to self
Known to others	Open area	Blind spot
Not known to others	Hidden area	Unknown

---

[6] The Johari window was developed by Joseph Luft (1916–2014) and Harrington Ingham (1916–1995) in 1955. It is a useful tool illustrating how people understand themselves and others. The Johari window comes handy when providing an easy to understand explanation why feedback is essential in the success of organizations.

The more you increase the size of the open area at the expense of the other three areas, the better you will develop as a professional.

Feedback should normally be a right, not a responsibility. A successful company cannot afford to work with employees having a low level of self-awareness and working on their facades to present themselves in a different way than what they are doing in practice.

Unfortunately, it is not always easy to exercise this right. Some companies are seemingly influenced by the mantra "no feedback is good feedback." If the feedback culture of your company is not satisfactory, I suggest the following action plan:

1. Say thank you often whenever someone deserves it.

2. Give feedback to your colleagues.

3. Observe how things change.

4. Ask for targeted feedback that is easy to give.

5. Establish a habit of exchanging feedback with your manager.

Say thank you whenever someone did something valuable that you appreciate. Tell them why their effort matters to you.

Give feedback. Develop a mentality of giving. Treat others in the way you want them to treat you. You may find out that similarly to you, others may be lacking feedback too. Even your managers appreciate feedback.

In order to avoid causing damage or ending up in your panic zone, start with positive feedback.

Observe how things change. Your peers may give you feedback more frequently. If nothing changes and people are intimidated by your feedback, something may be wrong with the company you are working for or with the way you formulate your feedback.

Occasionally, ask for targeted feedback. Once you finish a milestone, a project, or a hard task, feel free to ask for feedback. You can state that you are genuinely interested in improving your work, and therefore, you would be very grateful for advice in how to be more efficient.

Establish the habit of exchanging feedback with your manager. If regular feedback talks don't happen at your company, tell your leads assertively about your needs. Tell them that you have a hard time coping with the lack of regular feedback, as all the responsibility of optimizing your tasks is on you. You are certain that there are things you could do better, following the advice of other people.

The earlier you get to this stage, the better. Our goal is to establish regular feedback talks. Imagine working for a company with no feedback culture for a year. As soon as you contact your manager with the magic phrase "performance evaluation," chances are your manager will be confused. This is rarely the right time for you to talk about money. You want to be in a strong position. The path leading toward a strong position is through continuous action. If you work on the points you get from feedback, you will eventually be recognized as someone going the extra mile.

Dealing with Criticism

A lot of people take negative feedback personally. Don't be one of them. Your goal is to learn from all forms of feedback.

Whenever you get feedback, always ask yourself the following two questions:

- Who gave you feedback?
- What is the intention behind the feedback you got?

For instance, if a drunk person you had never met before told you that he hated you, you must be quite certain about multiple things. For instance, the person is not authentic in giving you advice. Second, his intentions were most likely not helpful. Most likely he manifested his own frustrations on the first person he saw. As a consequence, you can safely discard his opinion.

Similarly, if your colleague just got fired and he tells you that your work is the reason why he could not do his job, this statement does not reveal much about your skills either. This time, however, his opinion also acts as a warning.

There are two types of negative feedback:

- Constructive criticism
- Destructive criticism

Constructive criticism is nothing else but feedback that highlights one of your weaknesses for the purpose of you improving this area of your work.

Destructive criticism is something you get, because someone is either frustrated or has negative intentions with you. The purpose of destructive criticism is not that you learn from it and improve your work, but it is a personal attack.

In both cases, apply as much filtering as possible, ask less questions, and handle feedback as if it was constructive. Don't let your emotions take control of your actions. The last thing you need is an argument about destructive criticism.

If you ever give destructive criticism back, or you start arguing, eventually, you will figure out that you are no better than the other person. For this reason, view the discussion from a different perspective. People giving you destructive criticism will eventually figure out that they cannot fight against you, and for this reason, they will stop criticizing you.

Let criticism be constructive or destructive, you have to filter what applies to you and what doesn't. Your interest is to figure out as much about the feedback as possible.

Apply the principles of listening and reflecting. Use phrases like

- So what you are saying is …

- To check my understanding …

For instance, imagine you got the feedback that you are a horrible professional, because you are working with JavaScript, and JavaScript is a simple toy language that cannot be taken seriously. Therefore, you cannot be taken seriously either. You are imitating to be a software developer while you have no clue.

■ **Note** When receiving destructive criticism, chances are it does not pay off for you to learn from the feedback, because the feedback is disconnected from objective reality. Your job is to display traits of social intelligence to reach a consensus or at least avoid a pointless debate.

In this example, you could feed the feedback back by saying: "So you are saying JavaScript is a toy language that cannot be taken seriously and therefore I am not a serious professional? All right, this is also an opinion."

With this answer, you accomplish the following:

- You acknowledge that you have heard the message by feeding it back.

- You reach consensus with the other person by acknowledging that the message qualifies to be an opinion of someone.

- Once consensus is reached, there is nothing else to be discussed related to this topic.

You are not debating with the feedback. You do not state that you agree with it or disagree with it.

Unfortunately, the words are not enough. Depending on your tonality and physiology, this message may come across as playful, factual, robotic, ironic, sarcastic, defensive, passively aggressive, frustrated, angry, or even abusive. If you are not comfortable with this answer, you may choose to make your life easier and only say, "Thank you for your feedback." Unfortunately, this answer lacks the acknowledgement and the consensus elements, but you still have a good chance of avoiding a pointless debate.

You may decide on helping and coaching the other person instead of terminating the conversation. In this case, it may help if you point out some feelings you recognized:

- You seem worried about this one point. Tell me more about it, please!

- You sound a bit confused about the potential solutions to this problem. Shall we think about it and get back to this specific problem after research?

When a point needs clarification, feel free to ask:

- Could you please clarify what you mean by that?

- Please explain this thought in a bit more detail.

Our goal is twofold. First of all, we need to figure out as much information as possible, because even destructive criticism may contain some traces of constructive feedback. Second, we need to make sure we are actively listening by considering feedback.

A necessary condition of easing feedback is to give up on arguing. When receiving feedback, lower your defenses, and listen. Even if the feedback is wrong, you do not benefit from arguing that you are right.

When you receive feedback from your manager, say thank you after receiving it regardless of the underlying intent. Reiterate some points of the feedback, and promise your manager that you would work on those points and get back to him or her with your progress.

Whenever you talk about your own feelings, make sure you stay assertive. This means

- You communicate how you interpreted a statement.

- You communicate what feelings your interpretation triggered in you.

- You communicate a consequence of these thoughts and feelings, considering the point of view of both parties.

For instance, suppose your manager told you the following: "You are a very accurate problem solver. Regardless of the situation, you always find a perfect solution, and you hardly ever make any mistakes. For this reason, I encourage you to speed up a bit, even if it comes at the expense of making a bit more mistakes. Place trust in your colleagues; they will eventually catch your mistakes."

Suppose that you interpret the preceding message with your internal narration in the following way: "My manager is telling me that I am very slow. Is he thinking that I am useless and never get anything done? All my colleagues make a lot of mistakes, and I caught all their mistakes. How could I ever trust them? They make the mistakes, I am error-free, and they are the ones who get rewarded. This is unfair!" Do you have the right to formulate this internal narration? Absolutely!

Note   The internal narration associated with receiving feedback should be considered as a reification statement. Reification statements are facts. Debating how people should feel in your opinion is considered a violation of psychological boundaries.

Boundary violation occurs when you disown your own choices and blame someone else for a situation. People have the right to interpret feedback in any rational or irrational way. When you blame them for an interpretation that does not make sense to you, you violate the boundaries of the other person.

The tool that guards you from violating the boundaries of others is assertive communication, because it forces you to consider the interest of the other party. Assertive communication also helps you prevent others from violating your boundaries, because assertive communication represents your interest.

Let's answer our manager's feedback using passive/submissive, aggressive, and assertive communication.

The passive way is just to repeat and accept the statement: "Yes, it is possible that I am slower than what I should be. I will focus on productivity more in the future." The problem with the passive answer is that your internal narration will continue in the future. The passive answer does not represent your interest.

The aggressive answer is quite destructive, and this may block future feedback coming in your way:

- "Are you telling me that I am slow?"

- "No, I am not slow! Opposed to others, like you know who (!!!), I am not gambling with the resources of the company!"

Regardless of the answer itself, note that the aggressive approach attacks the person who had the intentions of trying to help you with feedback.

Let's see the assertive approach: "As you were giving me this feedback, I felt a bit frustrated for a moment, because first I understood it in a way that I was very slow. Thinking about the situation, I am aware that you have a good reason for giving me this feedback. There are also reasons why I haven't managed to speed up. I would like to share my reasons with you, so that we can draw some conclusions together in how to move forward."

The assertive approach clearly wins. Your manager will eventually help you settle your internal battle, and your internal narration on this subject will eventually stop. You put your feelings on the table, while you chose to exercise emotional control. You chose not to get offended, and you chose not to make a fast judgment on the feedback. There is also an action plan following up the main points of the feedback. This is the approach of a professional.

There are people out there who are not able to sleep because their internal narration about work never stops. Whenever you take a passive approach, you accumulate tension. The aggressive approach temporarily releases some of this tension in you, but it is not intended to provide you with a solution or long-term relief, because future frustrations and anger will accumulate more tension. Eventually, your perception gets distorted, and others will not be able to give you feedback.

The only win-win situation comes from the assertive approach. The assertive approach neutralizes tension for both parties.

■ **Note** An important element of receiving feedback is emotional control. You are responsible for your own emotions. Emotion theory distinguishes two components of emotion: the phenomenon and our interpretation. In general, events are neutral, and we often have no control over them. What we can influence is how we react to things that happen to us. If you get offended and become sad, angry, or fearful, it is your choice.

As I consider myself meritocratic in nature, I tend to value performance over other factors. Therefore, it is often evident to me that I need to point some things out and give feedback to others.

I am still working on my number one flaw at the time of writing this book, which is the lack of acknowledgement of destructive interpretations of feedback. My strategy of overcoming this flaw is reminding myself that not acknowledging the right to interpret an event in a destructive way is nothing else but a violation of the psychological boundaries of the other person. This way I gain emotional control over my tendencies of not acknowledging interpretations that violate my beliefs about meritocracy. This level of emotional control comes handy in conflict resolutions as well as in giving feedback.

Giving Feedback

The first chapters of the book *How to Win Friends and Influence People* by Dale Carnegie emphasize the fact that people want to be praised and people want to feel important. Furthermore, positive reinforcement is the only type of reinforcement that works.

The only way for people to see they are important is that you put their interest first. Try to be in their shoes, and ask yourself how they would appreciate your message.

Given that you care about your colleagues, they will understand your message. Giving feedback should not be characterized by euphemism or hiding the truth. Feedback is rather a tool to help someone get better.

Some tyrannical managers tend to propagate pressure downward. They tell their subordinates how unreliable, bad, and incompetent they are. In other words, they use fear to command others. This is wrong. No one will tolerate such behavior in the long run. Aggressive communication exploits other people, and they will do their best to avoid such interactions, even at the expense of resigning.

Giving great feedback is essential, yet it is very hard at the same time. If your feedback sounds positive and empty, most people will figure out that you are not honest.

Many people also know the praise-criticize-praise sandwich formula. They want to give constructive feedback, and they relax it by wrapping their constructive criticism with two praises. Although it is still more effective at the workplace than brutal criticism, this technique can also be slightly manipulative, when used wrongly.

I call a technique manipulative, when we are formulating our message such that we say things we don't genuinely want to say, just to trigger certain feelings in the other person.

For instance, the praise-criticize-praise sandwich is manipulative, whenever you include praises consciously, just for the sake of completing the praise-criticize-praise sandwich. People blindly following this method start with an empty praise, talk about the second part in depth from their own perspective, and finish with another empty praise. The same people wonder why their feedback is not effective and look for another Holy Grail pattern to follow. In reality, the Holy Grail they need is congruency and taking responsibility.

Praising, then criticizing, and then praising only works if the two praises are also honest and you would have mentioned these praises anyway.

Formulate feedback from the perspective of the other person. Don't be afraid of telling the truth in the right way. The trick is to find out how and when the other person is open to receiving feedback.

Negative feedback should not appear as criticism. You should help other people. For this reason, you should stay as factual and as helpful as possible. The following guidelines will help you:

1. **Collect facts**: Make sure your feedback has solid grounds. Whenever you have no evidence, you are just guessing. In this case, asking questions and informally talking about the topic is superior to giving feedback.

2. **Connect these facts with consequences**: Describe the effects and side effects of the area you are reviewing.

3. **Establish a common understanding**: Make sure both of you understand the situation in the same way and there are no pieces of information that would make you draw wrong conclusions.

4. **Agree on a desired outcome**: Ideally, the desired outcome should be SMART, that is, Specific, Measurable, Attainable, Realistic, and Time-based. Sometimes you may leave out some or all the letters. For instance, if someone has problems with their personal life, we show maximal understanding, and we are not going to handle the life of our colleague as a project. Giving your colleague a deadline to sort out personal matters makes no sense. In this case, the desired outcome is that we relax expectations until the problem is solved.

5. **Offer help and follow-up**: A meeting on its own is often not enough. Help is always appreciated.

This process does not say anything about your communication. Similarly to all areas of life, the way you formulate your message is a lot more important than the message itself. Your communicational tools are empathy, emotional intelligence, and assertive communication.

Whenever you are in someone else's shoes, you feel an empathetic connection. When formulating your message, always feel what the other person might feel, when reading or hearing your message out of context. This exercise will help you develop empathy.

You are assertive; you stand up for your rights and for other people's rights at the same time. You avoid being aggressive, and you also avoid passive actions by accepting what's wrong.

Let's compare feedback formulated in an aggressive manner and in an assertive manner using the five-step formula.

> "You failed to inform me three times on delays related to (...). Therefore, people working with you may think that you are not reliable. I guess you don't have any excuses for not informing me on these delays, right?"

There is not too much room for learning here. Either a debate is started, or the other person is trying to end the conversation here and now. Suppose that you conclude the feedback session in this manner:

> "From today onward, please (...). We will review all your projects in a month, and I will collect all delays on your end. If you have any questions, you know where you find me."

In this piece of feedback, the giver is straight up abusive. Therefore, the feedback is unlikely to become helpful. The first sentence already triggers defense mechanisms of the receiver. For this reason, the receiver may already check out mentally and retreat into his or her internal narration. Calling someone unreliable is a good recipe of destroying your relationship with your colleague. Talking about excuses just provokes arguments. At this stage, one of the following four things may happen:

1. Loss of your colleague's motivation.

2. Maximum defense, making the feedback message ineffective.

3. Your colleague attacks you, resulting in an argument.

4. Your message will be completely ignored. You might receive a thank you, and your colleague will avoid you in the future.

The giver of the feedback is thinking along the lines of monitoring and punishments. The only way to motivate another person is through making them want to cooperate with you, by

- Finding out and satisfying what they want

- Coming up with positive incentives

Let's formulate feedback on the same situation, using an assertive style:

> "I really appreciate that you are open for feedback, especially after such a hard project. I understand that in your situation, I might have not had time or mental capacity to inform other parties either. Let's use the situation to extract learnings out of it. Have you

> thought about what factors led to the delay? Is there
> anything you could have done to catch the delay earlier?"

This feedback opens a discussion to understand each other's point of view. You can conclude the feedback session in a positive manner.

> "To summarize the points, we have agreed on (...).
> This is a really nice outcome. Thank you for your
> thoughts. Hopefully, the new process will shift some
> extra, unnecessary responsibility away from you. In
> exchange, could you please confirm that you are fine
> with doing (...) as soon as you anticipate a delay?"

The second feedback seems to have created a win-win situation. Both parties are happy and satisfied. The feedback became a constructive discussion on how to solve the problem.

Notice that the aggressive method puts all the responsibility on the feedback taker. The assertive approach shares responsibility between the feedback giver and the feedback taker. This is the way to win friends and supporters in the long run.

Once I conducted a feedback talk, as I had the impression that a software developer was not owning his work well enough. In other words, he did the coding and relied on someone else to apply common sense and find errors in the solution. In software development, saying ready in a state where the code is still erroneous is like declaring that a surgery is ready while a wound is still open and bleeding. Let's call my colleague Frank.

If I had said, "Listen, Frank. You need to work on your accuracy, and you need to test your solution better," Frank would have either gone defensive, or he would have said, "Yes, sure, I will do it," without thinking seriously about it.

I was quite certain that Frank was also troubled a bit and there was a reason why his performance was slightly even worse than usual. I was also certain that poor guy was more unlucky than usual, as whatever could go wrong did go wrong.

Therefore, I started conducting the feedback by telling him that I would like to coach him a bit on how to increase his own value to the company. This message immediately grabbed his attention.

Then I asked him to evaluate different errors in the code that were deployed to the live system. He had to grade these mistakes from zero to ten. Zero meant that the mistake was inevitable; we cannot expect anyone to discover it. Ten meant that the mistake should have been discovered even by a trainee, even if the trainee just woke up and stared at the screen. I asked Frank to be brutally honest, as we were taking away learnings.

Interestingly enough, Frank discovered a couple of nines and tens, and his average score was around six. Frank already knew the outcome of the story and told me that this realization was quite shocking for him. I told him, "No one expects a zero-point score, and everyone has the right to make mistakes. The best developers may average one point; an average developer may average around three. And in your case, Frank, I know you are capable of scoring below three."

Then I asked him if he had any troubles, and I offered my help. Frank said he temporarily lost a bit of motivation, and indeed, we promised him an extra perk that was delayed due to administrative reasons outside our control. Furthermore, Frank was thinking about leaving the company, but eventually he decided on staying.

Eventually I told him that it is a huge asset to be reliable. Many developers never go the extra mile and never take responsibility in a sense that they own their tasks, not just solve them by barely meeting expectations. I gave him feedback that made sure he had something to focus on in case he wanted to get ahead. Frank's salary got increased by more than 30% within 1 year, without him applying any negotiation strategies.

We addressed all the points, and Frank not only got help in dealing with his issues, but he also managed to motivate himself to create extra value.

Giving feedback is therefore a very important tool in negotiating a better salary. Other people will try to learn your tricks of giving feedback, even though you have no tricks at all. All you need is to be assertive; make sure you figure out what people want and you talk about them, not about yourself.

Extra Missions

An important asset in negotiating a higher salary is to solve either the problems your managers are worrying about or the problems your company would buy. Remember money follows responsibility.

Suppose that you detect a major problem the company is facing and you know that you are capable of solving it. You approach your manager by saying that you are motivated in fixing this specific problem, because it would save a lot more money or bring a lot more profit to your company than the problems you are currently working on.

Extra missions should not replace your current responsibilities, but should rather come on top of them. In other words, instead of socializing for 2 out of 8 hours, you rather make a short-term sacrifice and make a difference.

In extreme cases, you can act as an intrapreneur and invent a solution to a problem your company is facing. If you create extra revenue or save costs, your company will most likely welcome your efforts. Some companies even have a fund for intrapreneurs launching internal projects. This means to you that you may get some budget and you may earn your salary in exchange for working on something you defined for yourself.

I have seen employees sell advanced trainings to the company they worked for. These people identified a need, made a lot of research, put in all the effort, and came up with a cutting-edge training. The company saved a lot of money with the training as well, so it was a win-win situation.

If you are really motivated in an extra mission, but you don't know what to do, become a mentor. Even people with little experience can mentor others, if they gain expertise in a field. You help others by mentoring and create value.

Exercises

■ **Note**   This is a section where exercises require you to take action in the real world. You may have brilliant ideas, but if you never start working on them, these ideas will be useless to you. Only action separates dreamers from achievers.

Communication exercises may be hard, especially for introverts. Whenever an exercise feels too hard for you, try with easier exercises first, until you are able to tackle harder challenges. If you need more help in this area, I can also refer you to my book *The Charismatic Coder* that helps you reach social flow and communicate in an authentic way effortlessly.

I cannot promise anything, but hard work in this section. The reward of your work is a massive improvement in your everyday life.

---

1. Name four things you can do right now to make yourself more effective at work. Start taking action by tackling these improvements one by one.

2. How often do you say thank you? Do you find yourself in situations when you could say thank you, but you don't? Do you also praise people beyond just saying thank you? Remind yourself to say thank you whenever it's appropriate.

3. Find an action to praise every day. If you are afraid of praising your colleagues, use the principles of identifying and eliminating limiting beliefs, and come up with a plan to stretch your comfort zone. For instance, you can first start praising your family members, then your best friend, then someone in a shop, and then a stranger. Once you built up momentum, you can start approaching some of your colleagues.

4. Practice active listening in everyday situations. Use formulas learned in this section to show you are interested in what your colleague is saying. Summarize, point out feelings, and ask for clarification, whenever it is appropriate.

5. Practice assertive communication both in and outside the workplace. Make sure you add value in the company with your communication style.

6. Practice assertive communication by giving constructive criticism in the next months. If you are not comfortable with this, try it outside the company first.

7. Ask one of your colleagues to give you honest feedback.

8. Ask your manager to give you feedback. Establish a monthly feedback channel. Take all learnings seriously, and improve your work.

Getting Promoted Without Negotiation

Goals of this section:

- Determine the tasks that get you promoted.

- Understand the importance of taking responsibility.

Believe it or not, in many companies, you can do miracles if you put enough effort into the right tasks at the right time. The right tasks are not the tasks everyone else is doing and may not even be the tasks you did in your past.

Your tasks can be placed in two buckets:

1. Tasks that you work on in order to avoid getting fired (duties)

2. Tasks that you work on in order to get promoted (opportunities)

In the first category, there are mostly tasks that you get from other people. Delivering the minimum in order to avoid getting fired is very dangerous in the long run, as you will miss out on getting better. It just takes an economic collapse to find yourself in deep trouble. Your salary will not grow much either.

Completely focusing on your duties and ignoring the second bucket may undermine your career growth. Focusing on mandatory low-level tasks only proves that your contribution is a great match for your current position. Promotions will be awarded to those who take the responsibility of going the extra mile.

It is your choice whether you prefer doing exactly what you are told to do or you apply common sense and develop problem-solving skills.

Economics is all about demand and supply. There are a lot of people who are not willing to take responsibility. These people earn less, as they are replaceable. You would be amazed how scarce of a resource professional business-savvy problem solving is. By taking responsibility for all aspects of your work, you can easily reach the top 5% of your field.

Whenever onboarding new software developers was one of my duties, I always emphasized the following message: "You were the one person chosen out of 20 applicants. The difference between you and everyone else is that you demonstrated a high level of problem-solving skills, cooperation, and a mature thought process. In this company, everyone is expected to take responsibility for their own work and apply common sense beyond just blindly interpreting requirements. Please remember to take care of your immediate surroundings. Communicate openly with other teams, and make sure you understand their work to the extent that you can cooperate with them without any external help."

I was privileged to work in a dream team, as the whole recruitment process was set up in order to find people who are willing to take a minimum level of responsibility.

Initially, it is hard to take decisions, as you feel some uncertainty inside you. However, once you pick up momentum, you will figure out that you own your work. This makes you happier, as you feel that your contribution matters.

As an exercise, go out after work one day, and take a bus or a metro ride. Start observing other people. Can you see happy, energized people who can't wait to change the world? Do you rather see slow, uncertain people, who barely have energy to reach home and turn on their TV or game console?

People who like their jobs are generally happier and more energetic. The path toward lasting happiness in employment is through giving value, making a difference, and taking ownership for your actions. Full ownership requires some form of responsibility on your shoulders.

If you are not used to it, responsibility comes with a high level of tension. As soon as you get to the next level, this tension will be gone. You will think of yourself as someone moving toward their goals.

Money or Responsibility?

Sometimes you are faced with the dilemma of choosing between more money and more responsibility.

For instance, you can choose between the following jobs:

- A very well-paid job at a company, where you will have to do the same, repeatable task until you resign. Your salary will still grow, as you will be 10–20% more productive with experience.

- A slightly lower-paid job, where you can gain a lot of experience. You can also expect to double your salary within a year.

Surprisingly, many people choose the first job and figure out only later that they wasted 1 or 2 years of their lives.

I personally faced the challenge of choosing between becoming a lead or becoming an expert. Many people want to be experts of their field. Yet, they give up their specialization in favor of becoming a leader. Given that these people are not passionate about leading, they will mostly be disappointed. The reason of their disappointment is that they expect earning more money, and they might even earn a fraction more by becoming a lead, but in the long run, they would have earned a lot more by staying an expert.

I chose a hybrid approach of becoming a tech-savvy manager. I am still in touch with technology, while I help wherever the company needs my judgment in improving processes.

In the 21st century, being an expert is a role that pays off. Salaries of experts are often higher than salaries of team leads.

In the long run, choosing more responsibility is almost always better. Money will eventually catch up with responsibility.

■ **Note** Competence is created through tackling challenges. Responsibility follows competence. Money follows responsibility.

A decision you might want to take today is to take every opportunity you can to do something useful. The self-esteem and confidence boost you get as a side effect will pay off.

First of all, if something is bothering you, fix it. Suggest innovative ideas that boost productivity. Be the pioneer of trying good ideas out in practice.

Solve tough problems on a regular basis. You will soon be known as the person to turn to, once a problem becomes complex. With the right attitude, you can keep yourself motivated, as solving hard problems may even become addictive.

Turning down promotions may pay off in the long run. Once I had a choice to apply for a product director position. It would have entailed a higher income and new connections. Unfortunately, this was not the area of my expertise; therefore, me taking this position would have capitalized on the Peter principle making my employers believe that my engineering competence easily translates to product competence. I worried about tricking my employer less than a potentially hostile threat from the perspective of my career: tricking myself.

While I have been doing software engineering in one form or another for over 20 years, my product management experience is limited. My point of view is biased from an engineering point of view. More importantly, I would miss engineering, and I would disrupt my growth trajectory. The slightly higher salary would not compensate me for these losses at the current stage of my career. This is why your career plan is important when evaluating your options.

Exercises

1. Who is involved in the decision-making process of your salary?

2. What challenges are these people facing? What can you do to undertake some of these challenges?

Summary

The key takeaway of this chapter is that assertiveness pays off. Representing your own interest while taking the interest of others into consideration is a necessary component of career progress.

As you take more responsibility, the importance of social intelligence increases in your career. Social intelligence comes with emotional self-regulation. While others get triggered by random events, you have the option to react to events in a mature way and learn from every situation.

Social intelligence also gives you the option to train your mind that there are more opportunities you can take than you would think. This implies that manipulation will not make sense for you, because the rewards will not be lucrative for you. Although you may conclude that manipulation does not pay off for you, it does not mean that you are vulnerable to getting manipulated by others. Knowing how to defend yourself is essential.

Rejecting manipulation means that you shift your actions from explicit influence toward implicit influence. Instead of bragging about yourself, you may choose to let others conclude and internally narrate in their minds how important your contribution is.

Regardless of whether you work for a startup or a corporation, you have a right to receive feedback on your work. This chapter highlights effective ways of giving and receiving feedback. Feedback also helps you discover growth opportunities and extra missions you can undertake.

Sometimes you may get promoted without you initiating it. Promotions follow taking responsibility for not only doing your job but creating innovative solutions that help your team.

If you take these steps seriously, chances are that you won't even need the next two chapters. In fact, you may end up in a situation where you have to say no to many opportunities to have enough time for the ones that are right for you. Fortunately, saying no is an assertive trait, which naturally develops with training your assertiveness.

In the next two chapters, we will continue building on solid foundations and learn how to negotiate as a professional.

Negotiating Raises and Promotions

Get Paid What You Deserve

This chapter gives you a framework about negotiating raises and promotions. We software developers often reach a point where we have to assert ourselves in order to get what we really want. This happens from time to time even if our managers back our progress up and have good intentions. Sometimes they are not aware of our progress.

Imagine that you are in charge of a department of 20 people. Would you be able to track everyone's progress? If you are one person out of 20, your future is mostly in your own hands.

A lot of developers think that their managers know every effort they do and reward them accordingly. When the rewards do not come, they become frustrated and wait even more. The frustrations will distort their image of the same managers who backed them up before. Then the time comes when they contact their managers about a performance review. How does this review usually play out?

© Zsolt Nagy 2019
Z. Nagy, *Soft Skills to Advance Your Developer Career*,
https://doi.org/10.1007/978-1-4842-5092-1_6

Not too well.

You may ask for a raise at the wrong time, or worse, you may burn bridges with your attitude. Your image in the company may change.

When you are under emotional influence, your decisions will be suboptimal. This is what we will fix in this chapter. You will get a full framework for targeting a raise or a promotion.

In the first section, you will understand how salaries are managed and what the company wants from you. You will learn how to time your request right and how to deliver extra value with your pitch.

The second section will teach you how to pitch your request. You will first create a salary research document to back up your claims with data. You will learn how to communicate your vision about your role in the company. At the end of the section, you will formulate an irresistible offer.

The third section is about the ins and outs of effective negotiation of raises and new positions. We will bust a lot of myths that hold you back from focusing on what really matters.

Once negotiation is over, the fourth section will show you how to show gratitude and how to follow up your negotiation with actions.

In the fifth section, I will give you some reasons why it does not make sense for many people to go out and get an external offer before asking for a raise. This will give you a lot of free time and energy to focus on your quest of getting a raise.

Understand Your Negotiation Partner

Goals of this section:

- Identify the details of how your company manages salaries.
- Develop strategies for targeting raises at the right time.
- Understand the interest of the company.

When negotiating a favorable outcome, you can maximize your own chances by stepping in the shoes of the other party. You can get valuable information, insight, and strategies if you know more about the company you are working with, the managers responsible for your progress, and future projects you may participate in. By proactively finding out how you can add the most value, your job at negotiating a raise will become easier than without this step.

Understand How Salaries Are Managed

In order to negotiate a raise, it is important to understand the forces that shape salaries. These forces consist of company procedures and the internal narration of decision makers.

Company procedures differ from company to company. Tech startups are more flexible than governmental institutions. Smaller companies tend to be more flexible than large corporations. On the other hand, large corporations are typically wealthier, resulting in higher average pay.

Companies tend to associate a salary range with each position they offer. The goal of the recruiter is to find someone suitable for their position, requiring a salary inside the range.

Typically, employees are the first to name the salary they are looking for. Sometimes, tough negotiators tend to force the company to reveal their range, and their interest is to hide their expectations until they demonstrate their skill set.

Your history determines your reputation. If you are a tough negotiator yourself, you have to offer something unique. Many people think they offer something unique, even though they are fully replaceable. Therefore, make sure other people get a good sample of your skills. This is why your personal brand and your interviewing performance matter.

In extreme cases, your recruiters may be your fans. Imagine you are a tech blogger and you are an expert of a skill a company needs. Your future interviewers may read your blog on a weekly basis, and their company may already be using an open source software you wrote. In this situation, you have the luxury of targeting a salary outside the range associated with your targeted position, and the company will most likely match your expectations.

In other cases, the recruitment team is not skilled enough, and they crack under pressure. One of my friends told me that he applied for an entry-level position, where the range was €40,000–45,000 per year, and this range also matched the industry standards. Based on false information and wrong calculations, my friend claimed that he was not willing to relocate below €52,000. He got the job. Later, the HR team disclosed that it was way beyond the norm, and they had to get an extra approval to pay this salary. However, once they got this approval, HR had no interest in not hiring my friend.

Getting Raises

Some consulting companies have a fixed salary plan. They have fixed positions, such as trainee consultants, junior consultants, consultants, managers, and partners. Your position determines your salary. The system is up or out. Rules are clear; there is no room for negotiation. In a company like this, you hardly have any time to negotiate anyway.

In other companies, including some famous financial institutions, there is one day for raises for the whole year. There are no exceptions. Even if you get promoted, you get to keep your salary for a whole year.

Governmental positions often come with a fixed salary. There is absolutely no room for negotiation. Take the offer, or leave it. Some of these companies do not offer any promotions either.

Your job is to figure out the rules. This can be done in multiple ways. You can network with your colleagues and openly exchange information with the right people. You can visit benchmark sites such as glassdoor.com or payscale.com. Alternatively, in some companies, it is possible to politely ask your manager, your department lead, the CEO, or the vice president about the internal rules. This requires a good relationship and a positive emotional balance. In other words, before asking such questions, you have to give, by continuously delivering quality work.

In some cases, your quest for a pay raise ends here. Take all the learnings from previous chapters; learn the skills you need to learn for a different job.

Fortunately, these cases are more and more scarce, as the interest of the business is to have options to keep their best people. Therefore, even if rules seem tough at your company, as a top performer and assertive negotiator, some doors tend to open up for you that are not available for the general public.

Whenever your salary is not significantly higher than the market value of your contribution, there is often room for negotiation.

Some companies have fixed raises, and they have a process for extraordinary raises. This process is often not convenient, and approvals have to be collected. If you are with a bureaucratic company, this process may be quite complex. Either way, your job is to figure out the obstacles and assertively follow up the progress of your quest. The good news is if your performance is exceptional, it is not impossible to get enough backup to make an extraordinary raise happen even if the general public thinks it is impossible.

Corporations tend to have rules, but these rules can be bent at the top. The other extreme case is when there are absolutely no rules. This happens in smaller companies, where you have limited budget, but flexibility is maximal. If your perceived status is very high and your role is unique enough, you might be able to dictate the price of your services. If you are one of the five experts in the world who is capable of handling the aftereffects of the explosion of a nuclear power plant and all five of you are needed for an emergency, you set your price.

The absence of rules does not mean that the sky is the limit. Treat this condition only beneficial for the absence of formalities, processes that bind your manager. You still have to do your best to increase your market value, and you still have to negotiate.

Most companies nominate a key person for handling the budget of a department. They are responsible for distributing raises based on the suggestions of leads and managers and some public or private heuristics. In extreme cases, even if the budget is spent, there is limited room for requesting additional budget in order to handle an emergency situation.

Let me clarify what an emergency situation means. Suppose you are a very important person, whose contribution is irreplaceable. If you leave, an important project of the company will be significantly delayed. Suppose you are earning €50,000 and another company is offering you €65,000. This is a take it or leave it situation for your company. If the company needs your services badly, they will be ready to pay you at least €62,000, if not €65,000, as they will still make money on your contribution and business continuity will not be under threat.

We will discuss the benefits and drawbacks of this approach. Waiting for a counteroffer upon receiving an external offer is a weak solution. There is almost always a cleaner solution that has an equal monetary expected return, with the side benefit of keeping your employer's trust.

Pay attention to one more important aspect: the financial year. Some companies get new budget before a new financial year. Time your request whenever you sense that there is some money left to accommodate your request. Your chances are the best a couple of months before the start of a financial year, as the budget owner has the time to use any unused money left from this year's budget or he can plan your raise in advance and possibly ask for more budget on time.

Some companies have a fixed month when they administrate all raises. If the financial year starts in January, it makes sense for the company to handle all raises in January and in July. Other companies operate with possible raises every quarter. Know these months, as worst case, you may end up with a waiting time of 5 months, unless you want to create a hard time for the budget owner.

Before we continue, let's summarize:

- Some companies operate with fixed salaries and fixed raises regardless of performance.

- Other companies operate with ranges.

- The more valuable your contribution, the more likely you can target a salary outside the range belonging to your position.

- Raises are usually bound by procedures and budgets.

- Emergency situations may justify overriding procedures and budgets, but causing an emergency comes with a cost.

- Know when raises are typically given. Know when the financial year starts.

- In extreme cases, there are no ranges, budgets, or procedures in place.

The Interest of the Company

Many people fail to consider their company's interest when planning the negotiation process. Especially narcissistic employees tend to think that no one else can match their performance and the company is significantly underpaying them.

Failing to understand what the company needs from you is a recipe for disaster.

Give something valuable in return for some benefits. In order to find out what your manager, budget owner, and company needs, you have to understand them on multiple levels.

Decisions are made by people. If you consider their interest, you will be more successful in the long run. Whenever you say something, other people start running an internal narration in their minds. This narration may include skepticism, doubt, warnings, or lack of interest. If you try to understand this narration, you will be able to avoid a lot of potential problems.

If you are looking for a counteroffer after revealing a higher offer, the narration of your manager may range from lack of interest to becoming desperate. Most likely, he may even consider that you are bluffing. This is one of the worst situations for you, as regardless of the outcome, the internal narration of your manager will hardly ever be positive.

Suppose that you approach your manager showing signs of frustration about your situation, blaming the company for lack of recognition. Depending on the personality of your manager, their internal narration will act against your message. What is going on? Why is this person so upset? Did I do anything bad? Can this person act in a professional way in the future? Has the relationship between the company and this employee been ruined already, or is there a way to save it? What other aspects may have triggered this behavior beyond money? Does this person know that these manners are not acceptable in the workplace? Shall I fire this person in order to minimize knock-on effects?

Your task is to make sure the internal narration of your manager backs up your argument. This is where emotional intelligence, empathy, and assertive communication are useful.

In some cases, your history gives you a big advantage. Suppose you are earning $80,000. You figure out that it is a good salary level in your city for your job at your experience level. However, you realize that salaries typically range between $60,000 and $100,000. You find one extremely high offer for $100,000 and a few offers for $90,000, and most of the offers are at or below $80,000.

Suppose that you have been giving maximum value since you joined the company. Suppose you always ask for feedback and always consider improving your work. You helped your manager multiple times by making sure you finished critical tasks on time, and once you gave them a suggestion to optimize some processes, saving a lot of manual work.

Your relationship with your manager is therefore very good. You show genuine interest on a regular basis, and you often talk informally with each other. One day, you mention that you need some help with orientation on your future. All you need to do is honestly tell your manager about your situation, making sure that your story is fully authentic and vulnerable. Some elements of such a message are as follows:

- The company is a great place to work for.

- It is my duty to inform you if I am facing challenges, so that we can work them out together.

- I am not going to let these challenges escalate to a point when I would threaten business continuity in any way.

- I respect you by telling you the truth.

- I am willing to solve any meaningful challenges that generate more value to the company, and you should not even worry about training me.

For instance, if you prefer keeping things simple, the following message will do:

> "I would like to find out how I can increase the value I give the company. I have some ideas, but I am not experienced enough to decide on this on my own. Could we brainstorm a bit together? I feel a bit stuck in my role at the moment, and I would like to know what the next step is for me to increase my earning ability."

This is a good conversation starter too. You make it clear that you want more money, but you are willing to give something in return. You take the interest of the company into consideration. The message is vulnerable as well. There is no explicit influence in the message, as you are not bragging about your past accomplishments.

Depending on the environment, the level of directness may be appropriate or too much. If directness is encouraged, you can even formulate a message like this:

> "Could we have a feedback session? I would like to figure out how I can be more valuable for the company. I feel that I am capable of earning more, and I would like to find out what I need to do, learn, or accomplish to reach the next step."

You are not hiding anything in this message either. Many people capitalize on feedback in a way that they are too shy to ask for more money, so they just ask for a feedback session. Then at the end of the feedback session, they say that actually they are quite frustrated with their situation and they would like to earn more money.

"As you said that I am doing great, I am worth more, right?" Well, wrong. No one likes to be trapped. This is nothing else but a trap that will shut down honest feedback loops in the long run, as your manager will be afraid of saying things that you can use against him to get another raise. Feedback is there for you to improve yourself and develop narrative influence. Feedback is not there for you to launch a rhetorical attack on the giver of the feedback to squeeze more money out of the company.

■ **Note** Do not use good feedback to trap your manager. The most extreme case I have heard of was when a manager told me his employee wanted to sue his company after he got fired, because the employee got good feedback a year before. Feedback is a snapshot of your performance and is intended to be used so that you improve your skills further. Feedback does not protect you from getting fired if you stop meeting expectations or do something illegal.

Exercises

1. Identify the procedures, budget constraints, and timing constraints of your company.

2. Gather all information about your role, your perceived value in the company, and industry standards. Take all benefits into consideration!

3. Decide if it is worth for you to pursue a raise.

Prepare and Practice Your Request

Goals of this section:

- Discover how much your services are worth in the market.

- Develop clarity about what you have to offer.

- Prepare an irresistible offer the company will have a hard time refusing.

If you think that your company is in a position to give you a raise, smooth execution and timing of your request is crucial. In this section, you will learn how to prepare your request.

Companies have a very hard time finding talent. Recruitment costs may reach five figures in dollars. Most of the time, it is important for your company to make you happy. This includes the degree of autonomy you are comfortable with, meaningful projects helping you unlock your potential, and more money.

In order to come up with the right pitch, you have to do three things. First, you have to research how much your services are worth on the job market. Then think about how you can contribute to the growth of the company. Come up with a vision of a compelling future that will give value to the company. Once you are clear about the value you provide and the raise you target, formulate an *irresistible offer*, and practice your request until you become comfortable with it.

Salary Research Document

Sometimes companies are biased when talking about salaries. The recruitment team often has many points of reference about people working in their company. If the average salary in the company for a given position is $65,000 and the market average is $72,000, the hiring managers may not even know that they are paying below the market average.

The smaller the company, the more flexible salaries are. In the case of big companies, policies about raises, starting salaries, and bonuses often tie the hands of the HR team.

It is your job to research your market value. The more accurate and the more unbiased your research is, the better. Include multiple sources in your research document. Some sources are as follows:

- PayScale[1] is one of the best resources you can get. Once you fill out their form on your current position, experience, qualifications, and skills, you will get a detailed salary report showing you the median salary, the top 20% mark, and the top 10% mark. You will also find out the percentile your salary is in.

- Glassdoor[2] gives you a lot of salaries. If your company is big enough or you are lucky, you may even find the salary of your colleagues.

- Stack Overflow Careers[3] gives a competitive advantage to jobs with a specified salary range. I have always found valuable and verifiable information there.

- Other job sites also have open salaries.

- Google is your friend. Especially for offers in the top 20 percentile, I tend to use Google as a job search site. For instance, if you want to find out if you can earn €80,000 as a Java developer in Munich, use the search expression "Java developer Munich 80000." At the time of writing this article, page 1 of Google gave me links to eight different job sites, with Java developer and senior Java developer positions that pay up to €80,000. Some of the indexed results may not be available anymore, but they can still be used in your salary research document

- Conversation with your colleagues. In some cultures, talking about salaries is a taboo topic. In others, people speak about salaries openly. Some companies force you to sign a clause in your contract forcing you to keep your salary a secret. Others publicly list everyone's salaries. Depending on the circumstances, I have always found partners for information exchange. The more you know about salary levels, the better your judgment is. After all,

---

[1] www.payscale.com
[2] www.glassdoor.com
[3] www.stackoverflow.com

you are not a CEO, and you are not working in HR or finance, so you are at a disadvantage when it comes to data available to you. Conversations about salary exchange generally do not hurt anyone, provided that you choose an emotionally intelligent and mature person for discussing salaries. Use these discussions only to formulate hypotheses about what is possible inside the company. Don't expose other people's salaries, and never formulate an argument around other people earning more than you.

■ **Note** Unfortunately, some employees have a distorted perception of salaries due to insecurities. Once they find out information about the salaries of other people, they compare themselves to them. This is when a developer with a fragile ego suffers, because they feel they are not enough, they are not worthy. Suffering can manifest in frustration, anger, resentment, being bitter, or sarcastic. If you feel this way, remember there is nothing wrong with you and you are not broken. You just haven't negotiated the terms for yourself that you think you deserve. I also encourage you to redo all exercises of Chapter 2 of this book.

Technically, you could talk to recruiters to get additional reference points. However, I advise against doing this, as you may easily end up with an offer even if you don't want to get one. Recruiters are very pushy, and they will do their best to earn their commission.

Once you finish gathering data, prepare a small salary research document. Use the PayScale template as an example, and include a verifiable link to it so that your employers will be able to regenerate the data if they choose to.

Include a couple of other links that you have found. These links should validate that all ranges of developer positions are available that serve as examples for the PayScale data.

Don't overdo it; you don't want to come up with 20 pages of references. A simple page containing PayScale data, some Glassdoor results, and some other job ads is more than enough.

It is likely that you won't need a salary research document at all. However, bring it with you once you make the pitch, so that you can hand it over if your lead wants you to.

■ **Note** The intention of the salary research document is not to create negotiation leverage to be used against your manager. A salary research document is intended to help your manager support you.

It is a successful strategy to be open about your research and sticking only to the necessary information. You may even mention numbers in your research once you get to it. It makes sense to hand over your research document by asking your lead if they are interested in using your research document to inform HR about market trends.

Vision

You have a chance to judge how much you are worth internally by considering feedback. No wonder why we focused on preparing feedback loops.

Knowing your percentile on PayScale helps you quantify how well you are currently recognized financially. Having done your salary research, you also know how much you should be earning based on feedback.

Your vision is about increasing your future value even further. Think about the challenges your employer is facing. Find out what you can do to improve your employer's situation.

This step is easier said than done. First of all, define the scope of your influence. There are often more than enough challenges in the field of software development. In some cases, you can also find some business connections between what you do and how the company makes money.

Your contribution can manifest in multiple ways. Consider these four categories in order of value to your employer:

1. You help your company make more money.

2. You help your company save costs.

3. You help your company reduce the risk of losing a lot of money.

4. You help your managers and your team become more effective, without specifying a quantifiable financial impact.

Even level 4 contribution is valuable. Find ways to make your team more effective. By thinking about your team instead of announcing your new prices in exchange for nothing, you are already doing better than 90% of other developers. Most people ask for a raise out of frustration, and they cannot think of anything but themselves. This attitude is not healthy.

■ **Note** We have already covered that some developers don't like to admit that business-savviness is an important skill. I claim being business-savvy is not only important, it is also necessary to grow beyond a certain level. Money comes from doing business. Therefore, if you want to earn more, it pays off to know how your contribution relates to the business.

Make sure you don't overdo any of your suggestions. You should keep good relationship with your team and never lobby for introducing burdens for the sake of your own progress.

Here are some examples for things you can introduce:

- Encourage knowledge transfer by volunteering to help others learn something important.

- Review the development process, and volunteer for extra involvement in trying out a change that helps your team.

- Figure out what tasks are hard for others, and take them over.

- Ask for an extra mission once you learn something the company needs. For instance, if you point out that there is no acceptance testing layer guarding your applications, you can volunteer to introduce acceptance testing.

- Introduce or improve coding standards.

The example about acceptance testing can become a level 3 contribution provided that you do a good job. Lack of end-to-end testing is often a major problem, as you have a hard time guaranteeing a high level of service.

Level 3 contribution is all about pointing out a potential risk. If the risk is worth fixing, your contribution will be valuable. However, no one likes to pay for things they don't want or don't need. Once you prove the problem is significant, you will be able to help your company eliminate the threat. For example:

- Improve unit, integration, and/or acceptance testing for a critical application.

- Forecast weak areas of your code based on past emergency situations.

- Benchmark the performance of your application, and identify the components of your software solution that are likely to become slow or unusable as the company grows.

- Point out a vulnerability in your system and propose fixing it. You would be amazed how many systems can still be hacked using SQL injection.

Level 2 contributions are the easiest to detect. If you can point out a change that saves costs to the company, you are on the right track. The developer mind often thinks about solutions that automatize manual work. All you need to do is quantify your efforts.

Automatize a manual process that requires a couple of hours of work every week. It is very easy to quantify the amount of money saved. Tasks that are important for the business are worth more than development tasks. If your idea saves your yearly salary, asking for a significant raise will be a no-brainer.

The only threat about level 2 contributions is that you don't want to automatize things that don't matter to the company.

As you reach level 1 contributions, miracles will start happening. You will be able to massively leverage your efforts. You will not only get a lot higher salary, but you will also be treated as someone who is suitable for a more responsible position. Level 1 contributions differ from company to company. Have lunch with people from other departments of your business to figure out what they need.

For instance, imagine that you develop software for a stock trader company and you are a hobby stock trader yourself. As you try out the services of your competitors, you find out a cool feature that lets you observe market trends more easily. You have a chance to use your domain-specific expertise to help your company introduce this feature and approach business to design a subscription model for it.

Alternatively, imagine that your company is using a service giving them bad user experience. If that service is easy enough to implement and host, you can either establish an open source or paid solution.

Level 1 contributions are hard and time consuming. However, some developers have been very successful with them. Some of my colleagues have launched their spin-off company on their product. It was hard work, but today, they don't have to worry about their own salaries anymore, as they pay themselves.

You should already have an idea about what matters to your company. This is what feedback loops are for. All you need to do is select some ways of contribution and make sure you execute them from start to finish.

Depending on the value you provide, set a target salary for yourself. If you are planning to provide a level 4, 3, or 2 contribution, you will most likely target an amount inside the payscale. Some level 1 contributions will identify you as an intrapreneur, and you will be able to target a salary that's higher than the top salary belonging to your position.

Irresistible Offer

You have to formulate a quick and efficient message to deliver your pitch. Make sure that you are straight to the point and you talk about value.

It is vital that your relationship should be good enough with your manager to talk about raises. Fortunately, the first five chapters set up a good and open relationship between you and your manager.

Practice your pitch at home, as a lot of money depends on it. You are not only fighting for an extra amount. You are also going to get a new base salary for your future raises.

Go back to the section on limiting beliefs if you are not comfortable with the discussion. Make sure you give yourself the permission to make mistakes and be cool. One way to train yourself is to record yourself speak in front of a camera. You will make mistakes, and your inner critique will make your life very hard. However, in the long run, you will allow yourself to make more and more mistakes without judging yourself. It is not a big deal to make mistakes, as mistakes make us human, and we tend to work with humans.

Giving yourself the permission to make mistakes is good for you. The reason is that a meeting on discussing your financial situation will almost never work according to the script you imagine. You cannot rule out the role of the other person in the discussion, and you have to be ready to handle the unexpected.

Accept that things will play out differently than you imagined. Accept that you may make mistakes. Accept that your lead may misunderstand your message and you will have to use different words to clarify the situation.

The most important element of formulating an irresistible offer is a captivating story. You will get a chance of telling a good story based on your past experience, highlighting the utility of your proposal. People relate to stories more than to numbers. Recall the section on creating your resume. It was also about telling a story.

The second most important element is that you should link your proposal to some business result that will be useful for your team, your manager, or your company.

Why Do I Have to Offer Something Beyond Asking for More Money?

You don't have to, but it makes perfect sense. First of all, you are able to charge more if you are special in some way. Second, you get more allies who may hold the keys to the budgets you are targeting. Third, you will build a professional relationship with your leads instead of capitalizing on it, and this will pay off in your career.

Sometimes there is a big gap between your market value and your salary. In this case, your claims are on solid grounds even without offering anything else. However, you will miss out on a lot of benefits by not offering anything.

In the unlikely case that you don't offer anything, make sure that you don't appear demotivated because of your salary situation.

Once I mentored a software developer who had sent an email to his lead stating that he wanted to earn more money, because he felt that he was losing motivation. He claimed the faster he got a raise, the faster he would regain his motivation. Obviously, this message was well hidden behind formalisms and clever language.

I am quite certain that this email undermined trust between him and his leads. Trust had to be repaired.

■ **Note** Your motivation is not a commodity. As soon as you try to leverage motivation, damage will be done, because you lose your attitude as a professional. Professionals are intrinsically motivated.

As a leader, you cannot expect that everyone behaves according to your professional code of conduct. Unfortunately, the majority of people rely on external motivation. It is a lot easier to play the trump card of losing motivation than taking responsibility for motivating ourselves.

It is important to know some people just won't accept the burden of taking responsibility and intrinsic motivation, because they cannot relate to delayed gratification. Furthermore, we are living in a world of consumption. Consumption is kept at a high level by offering choices to experience instant gratification. As life has become too comfortable, many people tend to lose their inner fire and just live a hedonistic lifestyle. While arguably there is not much wrong with a hedonistic lifestyle from a philosophical perspective, the trade-off is that the survival skills of the society start shrinking as we are encouraged to stay in our comfort zone.

The terms survival, responsibility, and intrinsic motivation may be scary for some people. If you are in a leadership position, you have to accept and live with these natural processes.

Earlier in life, I believed that I could just hire only those who take responsibility, and I would not have to face with these problems. Unfortunately, this is not how real life works.

■ **Note** A good leader can cooperate and motivate those who are not intrinsically motivated and can orchestrate and optimize work of those with different levels of professionalism and responsibility.

Therefore, as you add more value to the company, accept and respect those who contribute to the company, but don't share your views.

Exercises

1. Consider recording yourself as you talk about different topics on a weekly basis. Watch yourself afterward, and observe how your inner judge criticizes you. Do this exercise on a weekly basis, and watch how you give yourself more and more permission to be yourself.

2. Create a salary research document.

3. Design your pitch.

4. Validate your pitch in feedback sessions.

5. Schedule a meeting on reviewing your salary.

6. Practice your pitch before the meeting.

Secrets of Effective Negotiation

Goals of this section:

- Don't confuse yourself. Understand how you can stop creating obstacles for yourself.

- Learn how to formulate your message such that it will be easy for your manager to cooperate with you.

- Investigate the truth behind some what-if scenarios to get rid of your fears.

There is no good framework to handle every situation. However, there are some guidelines that help you.

First of all, don't confuse yourself! Most likely, your manager will understand your motivation.

If you have read some nonsense about looking like an alpha male or female during negotiation, make sure you don't lose your authenticity. First of all, you don't want to appear intimidating. Second, any technique that's a mere act will work against you. Don't forget that you are planning to negotiate with people you work with every day. Creating one persona for work and creating another persona to negotiate is rarely beneficial for you.

Several people told me, "Zsolt, you should read the book *Never Split the Difference*[4] by Chris Voss!" I did it, and I still say, as a software developer employee, negotiation is not where the big money is.

---

[4] Chris Voss, *Never Split the Difference: Negotiating as if Your Life Depended on It* (Harper Business, 2018).

If you run a business or you are self-employed, I understand that sometimes you need the edge, especially if you are up against business sharks or people who want to create win-lose situations for themselves at your expense. As an employee however, things are different. If your employer exploits you, as a professional, you have more than enough options to climb a ladder that's worth climbing.

■ **Note** There is a tendency from the end of software developers to blame their lack of success on not knowing how to negotiate. While negotiation comes handy, solely focusing on negotiation at the expense of performance may easily be harmful. Remember it is very easy to mentally outsource taking responsibility on some hyped magic pill techniques. While in some cases negotiation skills help you, they are not intended to provide you with easy solution where taking responsibility is the easiest solution.

Some people do succeed applying some fake techniques. However, the right question to ask is whether these people succeed because of or in spite of applying techniques at the cost of being authentic. Oftentimes, the answer is in spite of.

Most of the time, your manager will help you a lot more than you think. All you need to do is that you should clearly express your pitch. You have to mention that you are happy with your company and you imagine your future there. However, you are in trouble, as you have found evidence that the market pays significantly more for your services. It is all right to mention that you are not interviewing yet, as you place trust in the company. However, knowing yourself, you won't leave X amount of money on the table. Thinking about a potential raise made you think about how you can increase your own contribution to the company.

Then you can reveal your proposal. The frame of the discussion should be about the value of the contribution. You have to make sure your manager becomes satisfied once this part of the discussion ends.

Then you can ask for your manager's contribution for reaching the salary you targeted. You can tell him that you are ready to give him a salary research document if he wants to use it as proof. Assure your manager that it contains an unbiased aggregation of relevant data based on credible services. At this stage, your masterplan is done.

What if My Notice Period Is Very Short?

I have been in a situation personally, when I just had a couple of weeks of notice. It is very uncomfortable for you, especially if you develop fears and limiting beliefs telling you that you would get fired as a result of screwing up the negotiation.

No one will guarantee that you won't get fired. You may get fired even if you do a damn good job, just because of lack of funding. The question is how likely it is that you get fired just because of trying to negotiate a fair salary.

In reality, getting fired as a result of asking for a raise is a very unlikely event in general. The better you are, the more it costs for your company to replace you. We are talking about costs reaching five figures. It often costs your company more money to replace you than giving you the raise you asked for and financing it for a year.

My advice to you is to view your situation objectively. Is there a real reason why you feel uncomfortable? Is your relationship with your team and your manager good enough? Is there something you can improve?

Obviously, you can also go out and interview elsewhere. However, telling your manager that you have secured an offer elsewhere and you are asking for a $15,000 raise to stay is not the best thing that you can do from the perspective of your future relationship with your company, assuming the request comes out of the blue.

If there is a way, I normally prefer not interviewing elsewhere before making sure that I do everything to orchestrate cooperation on more solid grounds.

What if My Manager May Think I Am Trying to Take His or Her Position?

This scenario is very unlikely. However, I put this section in the book, because I have received this question more than once.

It is natural to be afraid of scenarios that don't exist. Ninety percent of our fears don't have a solid ground. Let's install some beliefs that make this event unlikely:

1. Cooperation always wins over competition in the long run. Your manager will realize that cooperating with you is mutually beneficial.

2. In today's job market, your earning ability does not directly depend on the number of people you supervise anymore. This was the earning model of the 1990s. Key skills trigger key rewards. A collection of independent key skills bring the company forward.

3. By the time you reach this chapter, you have received a lot of material on how to develop an open relationship with your manager based on trust. There are not too many people who master soft skills to this extent. You are open and transparent and get work done.

What if My Manager Won't Help Me?

Whatever you do, there will be cases when the relationship with your manager won't be good. We are not in a utopian world.

Negotiation in tough situations with tough managers is a difficult topic.

First of all, ask yourself the question: Does it make sense for you to spend another 1% or 2% of your life with people who can't stand you?

I won't tell you that you can achieve miracles with people who block your progress on purpose. It is unlikely that this happens, but if it does, note that fighting in an office environment will create a lose-lose relationship. Always stay professional.

It is not a big secret that our negotiation technique capitalizes on gaining trust of your leaders. If this trust cannot be gained with reasonable effort, you have to adjust your strategy. The principles still work. Even though it is a lot harder to come up with an irresistible offer, it is not impossible to create a win-win situation for everyone. You can still be open with your manager saying that you are going to ask for a raise. If he says no, you can contact another person who may listen to your ideas, or you can ask HR to ask how you can represent your interest. It is all right to be open about this.

In the unlikely case that your manager will start playing corporate games with you, stay on the professional side. If you have not done anything wrong, there is no reason to worry. A few people play games and use their authority, simply because they are not mature enough. These people often lose trust of their team and get fired. However, if you have no clear evidence that this is a likely scenario, don't confuse yourself with this section. It is a very unlikely scenario that your manager will start sabotaging you without any sign.

Worst case, make sure that you are prepared for some inconvenience. Documenting your decisions always comes handy, as you will always have enough evidence to combat unreasonable claims. Facts tend to defeat false claims. Professionalism tends to neutralize bad temper.

What if My Manager Tells Me There Is No Budget to Accommodate My Request?

Whatever happens, don't take things personal. Sometimes the hands of your manager are tied. Other times, the hands of your manager just appear to be tied. This is why you need to be assertive.

Based on your actions as a software professional, you should be known for being assertive at work. Therefore, during negotiation, you will be comfortable with representing your interest without fear while considering the interest of the company.

If your manager tells you that there is no budget, he should be aware of the circumstances. He should be aware of your market value. He should be aware of your value to your company. He is most likely aware of the cost of losing you.

If you have done a good job before negotiation, there will be an evident difference between your perceived value and your salary. Therefore, it will be very hard for your manager to say no. Your manager will know that it is not too hard for you to leave and earn a lot more money elsewhere.

As a consequence, your manager will most likely be open for cooperation. All you need to do is show understanding about the current budget situation and ask for a plan to reach your target salary.

Show that you are open for negotiating your level of contribution and you are always open for extra missions. However, showing that you settle for a lot less money in the long run makes little sense on your end.

The worst message that you can send out is that you understand the budget situation and you don't need that money that much. Staying humble is important in some situations; however, when giving a lot of value, you have to be able to accept the rewards that you deserve.

Your strategy is to keep your manager as honest as possible. You also provide honesty and transparency in exchange. This is a win-win situation.

The building blocks of your strategy will add up:

- You provide a lot of value.

- You are willing to provide even more value in the future.

- You strive to create win-win situations assertively.

- You are a hot asset on the job market.

- You are assertive enough to go after the conditions that you are looking for.

- You still give your company the priority to decide if they want to keep you.

These are the foundations for your irresistible offer.

Put yourself in the shoes of your manager. Without bias, how easy would it be for you to tell your colleague that there is no budget for the request, knowing that you had the opportunity to just reward your colleague with the raise? How much sense would it make for you to sabotage the progress of a person who could be your ally in moving the company forward? What are the chances that you decided on trying to fire your colleague? Would it even be possible?

Once you think these questions through, you will know how strong your position is. You don't need FBI hostage negotiation techniques to get a higher salary for yourself. Obviously, believing that all you need is some secret technique sells well, and it seems a lot less work than my approach. You may even reap some rewards with negotiation techniques, but they are just a fraction of what you can achieve by being assertive and striving for win-win situations.

Fulfillment is another factor. If you remember the section on hedonic and eudaimonic happiness in Chapter 2, more money may buy you some temporary forms of happiness. However, giving something in return and making a difference may give you happiness that lasts. Personal growth and making a difference in your sphere of influence results in eudaimonic happiness.

Exercise

1. Check each building block of your irresistible offer. Grade each building block from 0 to 10 based on how good your situation is. Then ask yourself what it takes to improve these points starting today. Once you are confident with these building blocks, you are on the right track to negotiate a raise for yourself.

Build Lasting Relationships by Showing Gratitude

Goals of this section:

- Understand the value of lasting professional relationships.

- Learn how to follow up negotiation with actions.

If your manager appears to be cooperating with you, show that you appreciate cooperation. Even if you left the company by accepting a different offer a couple of months later, showing gratitude would put you in a position that you would be able to work together with members of your current team one day. The emotional balance between you and your manager can stay positive regardless of the results of your negotiation.

The world is small. You never know when you will cooperate with your manager again. You never know when you will need a testimonial on what it was like to work with you. Furthermore, your life is far too short to burn bridges. Creating good relationships and building trust leads to win-win relationships.

What Happens After Negotiation?

Assuming that you are planning to provide more value, your relationship with the company will be stronger than ever before, despite them paying you more than before. In order to enforce this positive relationship, it is now your turn to go the extra mile and do your work.

Delivering more than what you promised is a foundation for future raises and promotions and will accelerate your progress. On a professional level, your skills will improve, making it possible for your managers to give you more responsibilities and more autonomy.

Document your accomplishments, and tell your lead what is done and when.

A lot of courses on salary negotiation never emphasize what happens after getting the raise. Psychologically, as soon as people get the desired external validation that their services are worth more, their confidence grows.

Staying humble is the key for raising your self-esteem. Know what you are worth, and show gratitude by going the extra mile.

Exercise

1. You only have one task to do after negotiation: come up with a plan for following up on your promise. Make sure you learn everything you need. Create and document your milestones, and inform your lead about your progress.

The Dark Side of Counteroffers

Goals of this section:

* Understand why leveraging offers results in lose-lose situations when you don't want to leave your employer.

Andy was frustrated because he did not get the financial recognition he thought he deserved. Therefore, he went out to the job market and started looking for companies. An agent contacted him. Andy first hesitated about naming his desired salary. The agent had a lot of experience in dealing with job applications, so he assured Andy that he was there to represent Andy's interest. Andy got relieved and told his agent about his salary expectations.

After reviewing some irrelevant offers, Andy ended up interviewing for a company he liked. He felt a bit bad, as he liked his current company more, despite the lower salary. He would have had no reason to leave his current employer if he had earned more. In fact, his current job was professionally more challenging.

After the interviews, Andy's agent asked him if there was any chance that his current employer countered the offer. Andy had to say that it was very unlikely, as they had been significantly underpaying him.

Andy got an offer that matched his expectations and felt thrilled about it. He saw a number materialize in front of him. He had been dreaming about this figure for months.

The next morning, Andy woke up feeling frightened, guilty, and depressed. He wanted to stay with his current employer. However, he had done nothing to make this happen.

Andy had to initiate an emergency meeting with his lead. Andy's lead expressed that he was already planning a raise for Andy. In addition, the company was very happy with Andy's work, and they were not in a position to replace him quickly.

While HR, the managing director, and Andy's lead were discussing the raise, the agent gave Andy many phone calls and Skype messages to push for the deal to happen. There was a lot of money at stake for the agent, as he may earn a five-figure sum upon clinching the deal. If Andy's current employer managed to keep him, the agent would earn nothing.

Note A recruiter typically earns 15–20% of your yearly salary if a company hires you. This is why some of them are quite pushy.

At the end of the day, Andy had to choose between two good offers. Regardless of his choice, he would earn a lot more money just by going out and getting an offer. On the surface, it seems that Andy can only win. Something is still wrong with this approach.

First of all, Andy showed lack of maturity and lack of professionalism. He successfully undermined his professional relationship with his company when it comes to getting promoted. Top positions require more autonomy, more trust, and more maturity.

Second, getting a raise at this expense may easily cost you a lot of money in the future. In the long run, the company may dampen your progress, as they will want to see a positive return on their investment of paying you more.

Third, your status will not change much after accepting the counteroffer. Make just one major mistake, or start underperforming, and your managers may easily start doubting you.

> ■ **Note** After a big raise, I have seen from time to time that people tend to relax, because they got what they were fighting for. Many managers are aware of this effect. Even those who are not aware will more likely notice if you are not overperforming anymore. Don't hold your performance back just because you got a raise, because it does not pay off.

Fourth, your self-esteem may shrink, and your guilt may not disappear for a while.

Note that leveraging offers is not the cleanest way of getting a raise. It is a weak action, caused by not taking responsibility for our own actions.

Some people equipped with narcissistic confidence keep telling themselves that they are worth more than what they are paid and the company is exploiting them. As a punishment, these people start applying elsewhere. Given that narcissistic confidence is an overcompensation for the lack of self-confidence, these people have to seek external validation to prove their value.

Once the offer is there, the developer comes back to their company. The company handles the emergency situation, extra budgets are accessed, and the pay raise is granted. The side effect of this process is that trust may never be restored.

Your work has to be extremely valuable in order to get away with an emergency raise in the future, regardless of whether you have an offer or not.

In this section, we will go over some common questions that arise when it comes to leveraging a counteroffer in specific situations. As every situation is different, you may choose to back yourself up with a counteroffer, even if you don't need it. This is sometimes important, because otherwise you would not have the courage to execute your strategy of asserting yourself. Therefore, even though counteroffers are often unnecessary, in some specific situations, they still come handy.

Are Counteroffers Always Harmful?

No. There are some situations when this is the only way to get a raise. However, ask yourself what this raise is worth for you. According to my experience, 90% of the time, it simply makes more sense to accept another offer and continue building your career elsewhere.

What if I Am on Probation?

You have read a lot of advice on how to build a professional relationship based on emotional intelligence, empathy, and assertive communication. You have had a chance to combat your fears and limiting beliefs. As long as you create an irresistible offer, the chances of getting fired are minimized.

If an external offer gives you the relief necessary to go through the process, it makes perfect sense for you to get an external offer. However, if you time your requests right, chances are that you won't even need to disclose that you have another offer. Successfully negotiating a raise during probation period without mentioning another offer will give you a lot of confidence.

This strategy works if you only use the offer as a possible safety net in the unlikely case that you get fired. As soon as you get evidence that you won't get fired, just continue building your relationship with your current employer.

What if I Don't Have Other Options?

You may be in a situation when you have a lot to lose by leaving your employer.

For instance, you may be in a small city, where there is one great company to work for, and your next option is several hundred kilometers away. Office romance could be another reason why you don't want to leave your current employer. Whatever your reasons are, sometimes you have a lot more to lose than in normal cases.

If you have a strong reason to stay with your employer, working on all components of the irresistible offer is even more important, as you are planning a significant part of your career with your current employer.

What if Corporate Bureaucracy Stands in My Way?

Andreas did an excellent job as in IT engineer at a multinational company. He was so busy working that he didn't pay attention to his colleagues seizing key positions.

As long as Andreas was busy, his manager was interested in keeping Andreas where he was, as he appeared to be a very efficient engineer.

Once Andreas started making it clear that he wanted to earn more money, his manager told him that the company procedures don't allow extraordinary raises, unless an emergency raise request procedure was launched. However, the manager was not interested in submitting this request, as he was about to leave his team.

Andreas got a new manager, who was very inexperienced with procedures. He also had no idea about the salary situation of Andreas. He had to start building trust from scratch, and he had to go the extra mile to ask his manager to go through the procedure to follow up the promise of the previous manager. This took a lot of time, as Andreas prioritized not causing too much inconvenience over his own interest.

What went wrong? Andreas was an excellent professional, who had a problem with asserting himself to the extent that is necessary to rise to the top. As soon as Andreas starts becoming assertive, he will have everything it takes to climb any corporate ladder quickly. The reason is that Andreas is a true giver. As soon as he specializes in focusing his efforts, he also reaps rewards in the process.

A necessary condition for being assertive about your raise is to get promises in writing, especially as soon as you know your manager may leave.

What if I Am in a Financial Emergency Situation?

Recall the safety instructions of your last flight: when the oxygen masks drop, help yourself before helping others.

There are no rules for handling an emergency. If you need an external offer, get it. However, note that sometimes your lead may help you without getting an external offer.

If you ask for a raise, go through the exact same process as without the emergency. Your financial situation should have nothing to do with the raise, except for the timing and urgency of your request. Don't leverage your financial trouble at the workplace, as it is unprofessional and counterproductive.

Make sure you minimize the damages. If you feel that you damaged your relationship with your employer, it is all right to tell your lead that you understand that the situation was extraordinary and you had a very strong reason to follow through with your plan. Desperate times call for desperate measures. In your situation, most people would have done everything to raise enough funds to solve a problem.

Summary

In this chapter, you have learned when and how to ask for raises and promotions, how to handle extreme situations, and what to do once you get the raise.

Be aware that your manager may have to execute processes that constrain the timing of your raises. Therefore, timing your request right pays off.

You also need to consider pitching your request properly. In this chapter, you saw how irresistible offers are created and how you can back up your research with data.

Once you have your pitch, you have to present it. Being authentic is often better than trying out a technique that's alien to you.

If you get what you wanted, show gratitude and don't hold your effort back.

Although it is often possible to get the raise you want without securing an external offer, in some cases, your best option is to leave your current employer and find a new opportunity. Therefore, in the next chapter, we will focus on finding a new job.

Get Your Dream Job

The Professional Software Developer's Guide to Successful Interviewing

After spending 10 years as a software developer, I have gained experience in both executing and going through interviewing processes. The reason why many applicants struggle is that they rarely get feedback they can use to improve their skills.

Even though I personally disagree with not giving feedback to applicants, I can understand that this is how the world works and this is what we have to accept. By asking for feedback, you maximize your chances of getting feedback.

The mental side of interviewing is also tough, because you will get rejected from time to time. Even the very best developers are rejected sometimes. We will focus on maximizing your chances of getting hired. We will focus on both the HR and the technical aspects of interviews. Even though this book is not about coding riddles,[1] I will go into details about the thought process of solving coding tasks and getting the best out of yourself.

---

[1] You can find some interviewing puzzles on http://zsoltnagy.eu and in my book *ES6 in Practice*.

© Zsolt Nagy 2019
Z. Nagy, *Soft Skills to Advance Your Developer Career*,
https://doi.org/10.1007/978-1-4842-5092-1_7

In the first section, you will read about two tech interviews. I have formed these two characters based on my experience having interviewed more than 500 candidates, emphasizing mindset and leaving out most of the subjective criteria I was looking for.

The second section is about doing your research. It is worth researching the target company before sending them your resume. I will give you examples on what to do and what not to do.

In the third section, we will cover interview situations from a nontechnical point of view. You will discover how important it is to come up with the right questions. You will also learn about the importance of integrity when it comes to answering questions.

In the fourth section, you will find out more about passing a technical interview. You will learn how to show your thought process and how to make your interviewers your allies. We will also discuss how to solve homework assignments and what you need to pay attention to when discussing your assignment in an interview situation.

The fifth section is on effective salary negotiation. You will get strategies backed by common sense.

The last two sections will give you two checklists on targeting a job with and without experience. These checklists act as a summary to verify what you have learned in the book.

The Story of Two Tech Interviews

Goals of this section:

- Compare and contrast the performance of a successful and an unsuccessful candidate.

- Learn how successful candidates make a lasting impression.

Two candidates, one goal: a software engineer job offer in a tech startup. Both candidates have had an equally convincing track record.

On the surface, there was not much difference between the two applicants, Matt and Dave. They even spent close to equal effort on preparing the application package. Yet, the contrast between them will soon become evident.

Matt crashed and burned during the interview. On his way home, he probably kept wondering why he always gets hard questions. He thought that everything went really well until he got a question that seemed so easy at first.

Dave not only got the job offer with ease, but he also got a higher starting salary than the maximum budgeted amount for the position. He also got a career path to become a lead developer. Dave was very happy with the offer and was proud that he passed an interview once again. On his way home though, he wondered which of his offers he should accept.

Why was Dave so much more successful than Matt? Observing the situation from different angles, at the end of this section, you will also conclude that success leaves clues. Sometimes it is evident why one candidate has a hundred times more chance of getting a job than another one, even if the successful candidate asks for a significantly higher salary.

The Application Process

For the sake of simplicity, we will focus on the following aspects of the application process:

1. Application package
2. HR interview
3. Homework assignment
4. Tech interview
5. Negotiation

We will analyze each aspect one by one and compare the performance of the two candidates.

All companies have different application procedures, and different interviewers look for different things. What works in one interview might not work in another one.

Application Package

Both Matt and Dave did a good enough job to get noticed. They demonstrated relevant experience in their résumés.

Matt had more experience, but less relevant experience. The only thing worrying me about Matt's application was his pride in counting his experience not only in years but also in months. Obviously, 10 years of experience is worth more on the market than 2 weeks. However, beyond a certain level of familiarity, the number of years of experience does not correlate with performance. There have been multiple studies about this phenomenon. For instance, in the book *Peopleware: Productive Projects and Teams,*[2] written by Tom DeMarco and Timothy Lister, Coding War games have been analyzed.

---

[2] Tom DeMarco and Tim Lister, *Peopleware: Productive Projects and Teams,* third edition (Addison Wesley, 2016).

People with 10 years of experience did not outperform their peers with 2 years' experience. Positive correlation between performance and familiarity only started forming below 6 months of experience.

On the surface, this seems to be a contradiction with respect to my article *"Design your Career – Become So Good They Can't Ignore You,"*[3] where I referenced the rule of 10,000 hours of deliberate practice.[4]

The paradox can be resolved by understanding what deliberate practice is. Deliberate practice requires effort with full focus. It is stretching experience, where you go outside your comfort zone and extend it. It is either practice spent in flow where you are so deeply immersed in an activity that you lose tracking time, or it is about mastering and perfecting a step of the activity deliberately until you reach the required standards.

This is why years of experience is not a good indicator for the amount of time spent on deliberate practice, leaving your comfort zone, and deep work. Some enthusiasts practice 4 hours a day, 300 days a year. In 2 years, they gain 2,400 hours of deliberate practice under their belt. A more relaxed developer may spend less than a hundred hours in deliberate practice per year.

Therefore, assuming that both Matt and Dave had a minimum amount of acceptable relevant experience in their specialization, I chose not to be impressed by years of experience as a metric. I kept Matt's month of experience counter in mind though, as it was clearly an out of line metric.

I recall another developer with 5 years of experience claim that he has had an experience of having written more than 100,000 lines of code throughout his career. This statement is equivalent to a bodybuilder saying that he has eaten 3 million kilocalories of food throughout his career.

If you mention the number of years of experience in too many places, chances are people interpret it as *explicit influence*. *Narrative influence* is a lot more fruitful, as you allow your recruiters to draw their own conclusion.

The resume of Dave was significantly more compact. He wrote everything that seemed relevant to us and left out the skills that didn't matter. After all, enumerating Microsoft Excel, a driver's license, and knowledge of Haskell did not increase Matt's chances. Less is often more.

The cover letter of Matt was all about bragging. It ended with the phrase "Because of my 6 years of experience, qualifications, and accomplishments, I believe I will be a valuable asset to your company." This was the one and only place where he mentioned our company in his cover letter.

---

[3] http://devcareermastery.com/design-your-career-become-so-good-they-cant-ignore-you
[4] For more reading on this subject, read the books *Outliers* by Malcolm Gladwell, *Mastery* by Robert Greene, and *So Good They Can't Ignore You* by Cal Newport.

Dave was a lot more easygoing, down to earth, and friendly. He was also quite informal. He expressed how happy he was when he read our job ad, as he always wanted to work in an environment where he could utilize his specialization to its limits. The cover letter sounded as if he genuinely cared about finding out whether we can work together.

I researched their LinkedIn recommendations and their GitHub profiles too. Both of them were cutting-edge. Matt seemed to have more dummy projects on GitHub. Dave collected several hundred stars on an open source repository, and it later turned out that a company was using this repository commercially.

Matt had a blog with five short articles. It is always great to have a blog even if it is not cutting-edge. The blog lacked quality though. It looked as though Matt had created a blog just to showcase that he had a blog. I personally still judged his blog as something positive.

Dave had a portfolio site, showcasing his open source contribution, including some demos. He also wrote about his mission in life. Very interesting read.

All in all, both Matt and Dave were interesting enough from a technical perspective. Dave clearly took the lead, but none of them have delivered anything so far in an interview situation.

Behavioral Interview

Matt continued applying explicit influence. He was emphasizing his own accomplishments at the expense of his team, sometimes voicing why each team he was working in was flawed and how he was superior. He didn't have many questions about the company or about the products the company is offering.

Matt didn't seem that interested in anything outside work. He also mentioned that his number one interest was to develop his own abilities.

Near the end of the interview, Matt started asking about overtime policies, the rules of booking leave, and benefits associated with the role. He said this information was important for him, as he had been stressed about getting too much work in the past and he was looking for a balanced life.

Matt didn't really bother revealing hardly anything about his personality. He kept his style professional. After all, this interview was more or less a formality for him. He did inquire about the next steps of the application process. At the end, both the HR manager and Matt left the interview room relieved.

At this point, some HR managers already show the door to even the best professionals, justified by lack of interest in the company. Matt did not make a lasting impression on the HR manager. His position was barely strong enough to get away with his attitude.

Dave came in 10 minutes late. He apologized for not being on time and assured the HR manager that he does not have a tendency of coming late. After apologizing, he didn't start explaining himself; he just sat down and felt comfortable.

Once I remember I was almost half an hour late from an interview due to a traffic accident. Things like this happen. The worst thing that can happen to you is losing focus. Finding excuses and explaining yourself is quite useless.

Dave kept his focus and started asking question after question on the company. The HR manager was genuinely surprised about the quality and quantity of questions. After a while, Dave showed understanding that the HR manager also wanted to ask questions.

There is a major difference between faking interest and showing genuine interest. Dave was curious about the company, his future position, and his colleagues, including the HR manager he was talking to. Instead of being pushy and trying to make an impression, Dave was trying his best to gather as much information as possible to decide if the job was the right fit for him. In some sense, he was the prize to be won, and companies were the ones bidding on his services.

After the interview, the company already had a great impression on Dave. When it came to Matt, he did the necessary minimum to pass all tests. His qualifications and experience showed he was the right fit, yet, he clearly did not go the extra mile.

Homework Assignment

Both candidates did an excellent job in developing all the requested features. Matt was very quick; he knew what he was doing. His total time spent on the task was 1 1/2 hours, measured from the download of the assignment. It took Dave 3 1/2 hours to complete the same task.

Matt understood the specification correctly and implemented all features adequately. The features were more or less robust; there were just a few small bugs here and there and one major bug that he overlooked.

Observing the code, Matt clearly had some difficulties with some of the more complex features. It looked as though he wrote some of his methods based on trial and error instead of planning them properly.

In Matt's homework, there was no generic commenting style in the application, and there was a clear absence of comments where they were needed.

Documentation was also fully missing. Even though it is not a requirement, documentation can sometimes save time even for developers. For instance, instead of implementing an obvious chunk of code handling an edge case, it is possible to just document the expected input format and list the full solution as an improvement suggestion that you are ready to complete upon request. The developer saves time, and the interviewer will know the absence of a feature was not due to negligence or lack of awareness.

Matt barely passed another hurdle once again. He completed all features quickly, so there was no reason to reject him. However, by now, both soft skill warnings and warnings on problem-solving skills have surfaced. These warnings will be examined during the second interview.

Dave failed to implement the last feature even though he was asked to complete all features. He wrote in the documentation that he omitted it due to lack of time, but he is willing to implement it in the upcoming version in case the rest of the code was not convincing.

The documentation of Dave was flawless. Installation instructions, assumptions, and software design considerations made it very clear that there was a smooth thought process behind the implementation.

Dave claimed that he practiced test-driven development during implementation. Indeed, all the tests showed intelligent design. I quickly recalled Matt did not even write any tests. Both Matt and Dave mentioned test-driven development in their skill set. Yet, Dave made it clear in his mission that he is always targeting 100% test coverage. Once I saw the task, I concluded these were not empty words. He provided test coverage even when he was not asked to do so.

All the implemented features were flawless. I spent 10 minutes trying to break the solution with irregular inputs. I tend to do this monkey test as a habit, and this skill has come very handy throughout my career. Astonishingly, I could only detect one small inconsistency. I could not even clearly call it a bug. I noted this inconsistency for future use, before examining the code more deeply.

The code itself was well organized and consistently commented. It was obvious for me to understand the code even without comments.

▨ **Note** "One difference between a smart programmer and a professional programmer is that the professional understands that clarity is king. Professionals use their powers for good and write code that others can understand."

—Robert C. Martin, *Clean Code*

After all the investigation, asking for the missing feature did not make much sense. Dave has demonstrated his approach toward craftsmanship already.

Tech Interview

The objective of Matt's tech interview was to determine if he was a good enough backup option in case Dave turned us down.

The objective of Dave's tech interview was rather to convince him that his skills would be utilized to their peak and he could gain more than enough career capital with us, meaning that he could grow faster with us than without us.

At this point, Matt clearly believed that he had nailed all hurdles so far. He arrived with full confidence.

I asked him if he had any questions. He had a few about our technological stack; then he asked about our development methodology. So far so good. Then he asked a question about overtime, followed by a question on performance review periods. His last question was about perks such as getting budget for a conference.

These questions revealed his selfish nature. First, he was looking for comfort; then he was looking for rewards. Being selfish is fine, as long as you can offer enough career capital in return. Some degree of selfishness is inevitable, both for your private life and for your professional life. Just recall the flight safety demonstrations you watch before your plane takes off: "Once the oxygen masks fall, help yourself before helping others."

Did Matt have enough career capital to offer? We soon found it out. I requested Matt to correct some of the smaller bugs he had in his homework. Instead of focusing on the solution, Matt focused on apologizing. He expressed how much he disliked these bugs and that he tended to catch such errors normally. I told him it was all right, as all of us make mistakes, and asked him to correct it in the code.

I showed him another small bug and then the most significant bug in his code. Apologies became more and more uncomfortable for him. This was when Matt realized that his chances were getting worse.

Unfortunately for Matt, the major bug was a hard nut, and it took him a lot of time and guidance to fix it. It was alarming to me that he did not log information that helped him; he rather treated a 30-line-long method as a black box and tested two modifications at once. He was in fact very lucky to have fixed the mistake without introducing two other bugs as a side effect. Having observed his thought process, I gave him one last chance to prove himself.

Matt received a technical riddle requiring nothing else but problem-solving skills. Matt had to understand a problem, formulate a solution plan, and implement his plan in about ten lines of code. Even though he had all resources available to him, he could have Googled anything he wanted, and he could have asked me questions. His problem-solving skills were simply not there. He started overcomplicating his solution, and he fell into his own trap.

It was clear to us that he would not have gotten the position even if Dave had not been hired. From the perspective of Matt, it may be hard to understand this decision.

It was a bit worrying to start the interview with Dave, as we had pressure to hire someone and we had no backup. After the introductory talk, Dave took control of the interview by kicking off a conversation about leadership style in the company, used technologies and methodologies, and he even showed interest in the history of the team.

I was convinced that his lexical knowledge was amazing, as it was developed out of interest.

We then moved on to talk about the homework assignment. I described the small bug in Dave's code. He said, "Nice catch," smiling. He jumped into his test suite, wrote a failing test, and corrected the code within a minute. Once his test passed, he showed the application in action. This process assured me that not only the bug was fixed but no other tests signaled any other failures after the change.

The problem-solving task was similarly easy for him. Dave explained his approach and translated his thought process into working code within 5 minutes.

The rest of the interview was a formality. We solved some problems together that I had encountered during work. Dave only got stuck with one problem, but even there, he demonstrated a brilliant thought process and contributed to the solution. Matt did not even reach this phase of the interview.

Negotiation Style

Matt did not have a chance to negotiate an offer. The only thing we know is that he was fishing for perks during the interview and he was trying his best to limit his own risks such as overtime. He often demonstrated his value based on explicit influence, expressing his thoughts on his own experience and knowledge. However, when it came to implicit influence, when it was time to shine, career capital was missing.

Dave was an assertive negotiator. He did not use techniques such as not revealing his salary until we expose our range. Instead, he exactly knew how much his services were worth and had a very good idea about our ranges too. He would have even been ready to name all the sources of his research in case his numbers hadn't matched ours.

All Dave said was "I am aware I am targeting a high salary, and it might be slightly outside your range. If you choose to hire me, I will make sure to do my best to justify the slightly higher costs."

As an added benefit, we talked about his ambitions during the interview. Dave expressed his interest in leading a small team. Although he did not get a guaranteed role, he got two promises: a performance evaluation within a year and a project he could take charge of. He also got full access to our leadership training.

Dave got the offer. Think about your own situation. Would you like to be in Matt's or in Dave's shoes? Think about one thing that you can do right now to improve your chances. Start developing momentum today by taking action.

Your Entry Ticket to the Interview

Goals of this section:

- Research the position and the company.
- Tailor and declutter your application package.

In this section, we will put everything you learned and created in Chapter 4 into the context of applying for a specific position. Remember you have to tailor your application based on the requirements of the position you are applying for.

Do Your Research

Some people spam a lot of companies regularly with one generic-purpose resume. A lot of developers think, "Senior Java developer position? No problem, I have experience in PHP, Python, and some JavaScript. Some companies will want me as I am experienced."

These developers often get low response rates for their application package. While reviewing resumes for years, the names of a couple of applicants became familiar to me, not because they were so good, but because they persistently sent me the same resume every 3 to 6 months. Whenever we defined a new role, regardless of the nature of the role, the application of a person I can clearly remember arrived quite soon. Their strategy must have been quite unsuccessful elsewhere too, as he was applying for years. People like him kept applying persistently, just in case someone once overlooked the evident lack of product-market-fit.

Some of these developers might go the extra mile to convince the recruitment team that they are the right fit.

One of the main fears of tech companies is that they hire someone who can't code. Therefore, many companies screen their applicants with techniques to collect evidence that their applicant has sufficient coding skills. If you don't show your expertise in the required programming language or technologies, chances are that you will only get a chance for the interview after everyone else with demonstrated relevant experience.

If you lack professional experience in a given technology, you can still get accepted for the first interview. Your career path should be sound. In your cover letter, it makes sense to *pre-frame* the situation. You can write that you have spent most of your career using technologies X and Y and that your skills are transferable to technology Z. As proof, attach an open source repository demonstrating that you have learned and used technology Z. Remember your learning plan; put it into action.

If you can't show any experience with a technology, don't get discouraged from applying. Your chances will be lower, but some companies are looking for people with problem-solving skills or a good cultural fit even if they are not specializing in some fields that are important for their role.

Note   Limiting beliefs of candidates often make them eliminate themselves before sending in their application. Many applicants think that they have to meet each and every point in the job description. In reality, this is not the case, and oftentimes the applicant who gets the job is not the one who meets most of the points in the job spec.

Do your best to tailor your application. Highlight skills that are similar to what the company is looking for, and make your cover letter stand out. Target the best possible cultural fit. You might be able to bond with the current employees of the company during the interview and convince them that you would be proficient in the technology they require by the time they get started. Although your chances will be lower than by meeting all criteria, it is also worth remembering that by not applying, your chance of getting hired is zero.

If you take an application seriously, research what the company is doing. Watch their career videos, and collect resources on the company.

Back in 2016, I did research on Spotify.[5] You can get information on their offices, how they work, and what values they find important. You can research people you would work with, visiting their LinkedIn profiles. Then you can find out more about their technology stack and read some tech blog articles directly from Spotify employees.[6] Do you want to learn more about the career steps[7]? It's on the blog. I then went on and found videos on the engineering culture of Spotify.[8] Having watched the video, I asked myself the question:

---

[5] https://spotifyjobs.com
[6] https://labs.spotify.com/
[7] https://labs.spotify.com/2016/02/15/spotify-technology-career-steps/
[8] https://vimeo.com/85490944

What has changed? Then I stumbled upon the post detailing why individual Objectives and Key Results (OKRs) are not used in Spotify anymore.[9] There you go, a great question to ask in the interview. If you research the recruitment process of Amazon, you can find even more information.

You can go a long way researching some companies. You may reject some companies you don't want to work for in case you find really out of line information. Best case, you will be so enthusiastic about meeting them that you will create an excellent interviewing atmosphere.

I highly recommend using Glassdoor[10] for researching what it is like to work with a company. Be careful though. Some people may leave great companies in a bad state of mind. These people skew their reviews to the negative side. Some companies encourage positive reviews or even fake them, to look better in the eyes of candidates. Therefore, I suggest taking an objective look at the reviews and extracting information that's useful for you.

Don't tell your interviewers that you have made a research on them on Glassdoor. It is sensitive information. In the unlikely case they asked this question, you can reply honestly that obviously, you used Glassdoor as a resource. However, trying to connect with the company by asking questions about Glassdoor reviews makes little sense on your end.

Your research also includes Googling about your interview experience. Glassdoor has an interview section, and Google gives you a lot of other secrets. Even though some information will be outdated, you may still get some reference points to do your best in the interview. Please don't abuse this information to get hired without competence; it will be a lose-lose situation for everyone. However, it is logical that you may use information you can find from legitimate sources.

Tailor Your Application

I know some people say that you have to hire a professional resume writer and spend $500–$800 on a cutting-edge resume. You may gain valuable experience in doing so, as it may be a great learning experience for you. However, this is not the most cost-effective solution to improve your chances of getting hired.

Chapter 4 was also a learning experience for you, and you have everything you need to create a professional resume.

---

[9] https://hrblog.spotify.com/2016/09/27/keeping-your-balance-between-chaos-and-structure/
[10] www.glassdoor.com

Tailoring your application is the task that brings you the most results. Using a professionally written resume for the purpose of mailing it to every company you can find is not likely to work for you.

Collect which competences your employer is looking for, and review your resume with the eyes of the company. If a part of your resume is not important for the company, remove it, or make it less significant. Highlight skills your employers are looking for.

Your resume is a one- to three-page ad that should attract the attention of your employers. Make it stand out, and make sure you find the right message. If you do your research well enough, you will know what a specific company is interested in. Use this knowledge, and send the right application package to the company.

Exercise

1.   Go back to Chapter 4. Reread the first two sections.

2.   Then find a company you would like to research. Find out everything about the company that's relevant to you from the perspective of creating your application package. Then open your resume and your cover letter template. Make changes to it based on your research to make it more relevant to the company you just researched.

3.   You might want to consider writing a cover letter from scratch. This makes perfect sense for your first two to three applications so that you gain more experience.

Acing the Behavioral Interview

Goals of this section:

•   Learn how to make a good first impression.

•   Learn why it is important to ace the first interview instead of just passing it.

•   Learn how to answer to some typical interview questions.

•   Find out what the right questions are that you should ask during the first interview.

Many software developers treat nontechnical interviews as the necessary evil. If you don't make an effort, you will leave a lot of money and opportunities on the table.

The nontechnical interview is always more important than it seems. This interview alone may have an impact on

- The difficulty level of the technical interview

- Your salary range

- Whether you are treated as a commodity or whether you can offer anything special that makes you stand out

- Whether you are a good cultural fit

- Your future positions, that is, if you will target a lead or an expert position later

Who Are My Interviewers?

Your interviewers may either be members of the HR or internal recruitment team, or hold key technical or nontechnical positions in the company.

Even though you are interviewed as a software developer, you might get a chance to meet a product manager, a project manager, a vice president, an architect, the head of UX, a software development team lead, the CTO, or even the CEO. In small startups, for instance, it is very likely that you are interviewed by the CEO or the CTO.

In extreme cases, some companies outsource recruiting to an external recruitment team. Most of the time, you still meet a key official of the company sooner or later, before getting hired.

If HR or an internal recruiter is present in the interview, make sure you treat her with equal respect as you would treat a technical person.

■ **Note** The number one mistake I saw in interviews is that some software developers show their superiority and show lack of respect and patience toward HR.

As your interviewer was chosen for a reason, it is likely that even a genius gets eliminated by showing lack of respect.

Once I heard that in an interview, the candidate only held eye contact with the technical interviewer, ignoring the recruiter. We will never know if this was an unconscious form of behavior. However, after the interview, the recruiter pointed out that she was ignored, and the technical interviewer started thinking about clear lack of soft skills too.

How Many People Interview Me?

Usually, one to three people interview you. There is typically one person representing HR and recruitment and up to two experts, leads, or managers.

Some companies, especially startups, tend to exclude HR and recruitment from the interview process. As long as they are small, they prefer collecting hands-on experience about the people they work with.

Other companies invite a lot of candidates and organize a full day for recruitment. This creates interesting dynamics, as you may meet a board of four to five people, and you have to cooperate with other candidates in solving and presenting tasks.

As a candidate, the most extreme first interview I have ever had was with four technical experts. Even though they worked in a corporate environment, they tried to create a relaxed atmosphere. However, connecting with all four of them was a tough job. When you are in a situation like this, don't focus on pleasing everyone. Just be polite and respectful, hold eye contact with everyone, and continue the interview as if you only had one or two interviewers.

When it comes to eye contact, some interviewers may sit next to you. This makes eye contact very uncomfortable, as you may have to turn up to 180 degrees while talking. It is all right to tell the person sitting next to you that you would like to suggest a different seating arrangement, as it is not comfortable for you to keep turning left and right, and it is not an option for you to ignore any of your interviewers. You set the frame, and your interviewers will cooperate with you.

The inner game of dealing with multiple interviewers is all about believing that you are the prize to be won, and the larger your audience, the more chance you have to really impress someone.

Anything can happen during an interview. Once a small startup invited me for an informal first interview. I got the chance to meet the team. During the meeting, the investor and the CEO showed up, and they assembled a 15-minute-long ad hoc meeting on software development matters. It was awkward for me to sit in there as an outsider. Then I met the investor, the CEO, the CTO, and some developers.

The more interviews you attend, the more comfortable you will become with extreme situations. Practice makes perfect.

Storytelling Frameworks

Many companies give you hints about an expected way to answer behavioral interview questions.

You might have heard of the STAR, STARR, and SOARA techniques by Hagymas Laszlo and Alexander Botos. These are as follows:

- **STAR**: Situation, Task, Action, Result
- **STARR**: Situation, Task, Action, Result, Reflection
- **SOARA**: Situation, Objective, Action, Result, Aftermath

These techniques give you a framework to tell a story. There needs to be an intriguing challenge that hooks the audience. You are the hero of the story who got the objective of accomplishing something facing some adversity. Suspense is created during the Action part. The actions you take should present a nonobvious path toward the desired result. Then you can choose to reflect on the story or not, depending on whether your audience can draw conclusions themselves.

A common mistake in telling stories is to add so many loose tangents into the story that makes it hard for the audience to follow the main plot and makes it impossible for them to draw the conclusion you want.

Another common mistake is when a candidate applies the storytelling frameworks to tell something obvious where there is absolutely nothing convincing.

Practicing Behavioral Interview Questions

I encourage you to read the following questions and answer them out loud. Record your answers with your phone, and listen to them afterward.

If you like challenges, record yourself with your laptop or phone camera, and watch yourself. This will not be a comfortable feeling for you initially, but your efforts will be worth it once you realize how much easier the real interview will become for you.

As you do this exercise, you will be in *training mindset*.[11] You will have the chance to analyze your approach and make changes.

---

[11] Credits go to John Eliot, PhD (www.overachievement.com/), for introducing the phrases training mindset and trusting mindset. During training mindset, your task is to practice different aspects of your performance to approach perfection. Without a conscious effort, your development is limited. During the act itself when peak performance is required, any conscious effort on observing yourself decreases your performance. Your best option is to trust your abilities.

Once you attend your interview, you will be in *trusting mindset*. All you need to do is trust your abilities and put your improved skills into practice. This time, you will have no time to analyze and fine-tune your approach anymore; therefore, you have to do your homework in advance.

We will simulate an interview. Before the interview, do your research on a specific company, and pretend you are in an interview with them.

I will give you all the questions at once. Do the exercise; then read my comments on the questions. Remember there is no single right answer. Your message should be congruent and sincere, and you should sometimes focus on avoiding some traps.

1. Please tell us a bit about yourself.

2. Are you interviewing with other companies?

3. Where did you see the position advertised?

4. Why do you want to work with us?

5. What do you know about our company?

6. Which was your favorite project/role?

7. Why are you leaving your current job?

8. What type of work environment do you prefer?

9. How would your colleagues describe you?

10. What are your hobbies and interests?

11. Do you imagine yourself as a technical expert or in a managerial role in 3 years?

We will use these questions to cover the basics of the interview. Obviously, there are some hard and unfair questions out there, but you don't have to prepare for every single situation.

If you are asked an illegal or unethical question such as your religion, sexual orientation, or whether you expect a baby any time soon, you can make it clear that you consider questions like this out of scope for the interview. You can only do this if you didn't answer any illegal questions before. Imagine answering three illegal questions and turning down the fourth one. It is almost as revealing as an answer, because you had a reason to cooperate in three cases and you have something against a fourth question. Remember assertiveness is all about defending your boundaries.

Take it easy about not answering a question you find illegal, as if nothing special happened. If the interviewer insists in getting an answer to an illegal question, my personal policy is that I politely tell them that based on the interview experience, I am not interested in the position anymore.

You may leave a backdoor open by softening your response, making a conditional statement, such as "If the application process requires me to answer questions prohibited by law, I will withdraw my application. Please let me know if we should terminate the interview or you are fine with continuing the interview without this question."

Let's get to dissecting the eleven questions.

1. Please tell us a bit about yourself!

This question is not about the story of your life, but it still includes storytelling.

Tell one or two interesting stories about your career. Focus on stories that are relevant to the position you are applying for.

If you have trouble with selecting a story, start with the present. What is your current position? Then shift to the story of an accomplishment in the past. What challenges have you overcome? What results have you delivered lately?

If you wrote a resume based on the principles of storytelling, all you need to do is elaborate a story or two and make it more energetic. You can conclude your story with a logical consequence of your skills and experience and forecast what you are looking for in the future.

2. Are you interviewing with other companies?

You don't have to disclose the companies you are interviewing with. However, it is common interest that you share your progress.

Some companies want to provide you with a good interview experience, meaning that they are sometimes flexible in scheduling your interviews according to your needs. This happens with many startups.

Other companies are less flexible. Some larger companies, for instance, have a fixed interview process.

One more thing to consider when answering this question is that you are the prize to be won. Of course, you are interviewing with others; it is natural.

As you qualify the companies you want to work with, it makes sense to interview with a few companies only. You can mention that you are only focusing on the best companies and you are not in final stages with anyone else yet.

If you are not interviewing with anyone else yet, it does not make much sense to fake it. You can just say that you most likely will contact a few companies, but you have just started interviewing, so this is your first interview.

3. Where did you see the position advertised?

The worst thing you can say is "I don't know. I have seen so many sources lately and sent out so many applications that I can't recall whether it was LinkedIn, Stack Overflow, or another job site."

Always be ready to name a source, where the company is advertised.

You should be straight to the point. This is not the right place to tell your employer about how much you liked the job ad. It makes sense for you to elaborate on your research on the company, because it signals that you put in some effort to understand the position you are applying for. If you were referred by a recruiter, tell the name of the recruiter. If you were referred by a current employee of the company, it is fine to say who your referrer was.

4. Why do you want to work with us?

This is a hard question, because you should show genuine excitement while keeping your prize status.

Your research about the company is very important. Do your homework and record a couple of things that make you curious about the company.

Many companies shoot a video to advertise themselves for recruitment purposes. It is easy to relate to the video. Alternatively, you can read their blog, check out their product, or attend their meetup.

It is inevitable that you will find some topics you can relate to. Filter these topics from the perspective of your skill set. Then tell a story.

Make sure you don't capitalize on something you are currently missing on that sheds bad light on your current or past employers.

For instance, you can formulate a statement like this:

> "In the last 3 years, I worked in a small startup, where we had ad hoc processes to react to the day-to-day challenges of the working environment. This was a great experience for me, as I grew a lot reacting to these challenges as a fresh graduate. I also got a chance to establish some new processes, because this was my specialization during my MsC studies. However, I would like to see a larger company now from the inside, because we are not at that stage right now as your company. I want to see how I can add value in an established organization with well-managed quality control. When I watched your video about the IT department, I saw that your company is at the stage of maturity that I am looking for."

This story is a lot more congruent than saying that you currently live in India and you would like to live in Europe. This message only reveals that the city and the company do not matter to you, just the continent.

Note There are two types of motivation: *moving toward* and *moving away from.* The problem with "moving away from" motivation is that the candidate does not choose a company. The candidate just plans their escape. This is not ideal, because interviewers are looking for skilled and motivated candidates. As it always takes two to tango, candidates who escape from another company likely have a character trait that may cause a problem in the new company as well. Therefore, people with "moving away from" motivation are less likely to get hired than people with "moving toward motivation."

---

5. What do you know about our company?

It is time to put your research into context. Express anything you are passionate about. For instance, if you like music yourself and you are interviewing for SoundCloud or Spotify, you have an easy job. If you used to trade with stocks and you are interviewing for a company creating real-time stock trader platforms, it is easy for you to connect with the employees of the company.

What happens if the company's products and services are not related to you and your life? You can still relate to the professionalism, technological challenges, technological stack, and the company culture.

Hiring managers would like to see if you can represent their company assuming they hire you. They would also like to see that you take your potential employer seriously by researching them.

If you run out of things to say, it makes sense to connect with your interviewers by telling them what you are surprised about, and you can ask a question or two in the meantime.

6. Which was your favorite project/role?

Your interviewers would like to hear that you are passionate about your job. The best way to showcase your passion is to talk about your past challenges.

This is an easy question if you are experienced and you have seen a couple of projects. Maybe you worked with a great team once. Or you had the chance to take full ownership for the project from start to finish, influencing the development methodology, the technological stack, and even the team. Alternatively, you can talk about overcoming major obstacles and delivering results you never thought were possible.

This is your chance to develop narrative influence, by telling a story about your strengths. If you want to be an expert, your favorite project should be one, where you had to do in-depth research and you had to be involved in key technological decisions. If you were a lead, your favorite project is most likely one, where results were due to smooth team effort.

7. Why are you leaving your current employer?

This question is not as complicated as it looks like. If you have done a good job with the previous questions, it is evident that you are looking for special types of challenges.

Never trash-talk your current employer. Always be grateful for the opportunities you got, even if you didn't receive the treatment you were looking for.

If you had conflicts with your employer, don't be negative about them. Every coin has two sides, and every conflict can be interpreted in multiple ways. Don't seek for external validation in a job interview. If you are not happy with your current employer, obviously, you are looking for more challenges.

There is a very thin line between right and wrong. If you want to be transparent, you can also say that you are not happy with your current employer, and it is not their fault. You just need different working conditions and professional relationships.

A variation of this question is about telling a story about an event when you disagreed with your employer, but you still had to do something.

8. What type of work environment do you prefer?

Connect the question to the research you made about the company. Some things must be important to you.

For instance, if you are used to daily scrum meetings and a full day of meetings every second week, chances are that you prefer well-established processes to ad hoc reactions. If trust and autonomy is important to you, it is all right to announce it. Do you prefer formal relationships or people who are down to earth and approachable? These are all important factors both for you and for the company.

Don't be afraid of formulating an opinion. Three things can happen afterward.

Your interviewers can recognize that they can provide you with what you are looking for. This leads them to proudly tell you that your preferences will be met.

Alternatively, the company may admit that they work in a different way. Especially in the case of startups, some procedures are not yet in place. This may lead to a discussion on whether they can meet your needs.

Some companies may sugarcoat or hide their point of view on your answer and move on.

Either way, formulating your opinion instead of seeking validation of your interviewers is beneficial. You stay the prize to be won.

9. How would your colleagues describe you?

There are multiple ways to approach this question.

First of all, if you have some LinkedIn recommendations, you can use it and tell your interviewers a bit more about your relationship with your past colleagues.

You can also refer to real praises that you got, let it be an informal meeting congratulating your team on an accomplishment or a performance review.

Alternatively, focus on a couple of your strengths, and put them into context. Formulate a story, and make sure that instead of bragging, you tell your interviewers an interesting story.

It is all right to confess some of your weaknesses as long as you have a good story and these weaknesses are not deal breakers for the job.

If you were a freelancer, you must have had clients. Talk about your relationship with your clients.

10. What are your hobbies and interests?

It is hard to go wrong with this question. If your answer is not empty and your answer is politically correct, you should be fine.

Your interviewers want to see what you are like as a person. They want to be able to relate to you in some way, and hobbies are important.

Faking hobbies does not make much sense. Some people recommend that you find out who your interviewers are, what they like, and pretend that you like the same things. Remember once you are hired, you will work with these people.

If you say that you like browsing the Internet and watching TV, your image will not be that good. If you want to be discarded quickly, combine this answer with the story that you would like to move to Canada and the job or the city does not really matter to you. All that matters to you is that you get a good Internet connection and great TV channels. Your interviewers will not likely believe that you will be a good cultural fit.

11. Do you imagine yourself as a technical expert or in a managerial role in 3 years?

If you have done your homework, it means that you have designed a career path for yourself. In this question, you get a chance to express what you are looking for.

Some people have problems with stating that they actually want to become leaders. This could be because of lack of self-esteem, or because they are afraid that their interviewer would sabotage their progress.

Nothing is further from the truth. Good leaders are in high demand. Many companies prefer filling lead positions from within, and you are offering them the opportunity to take your cooperation to the next level if both parties see it fit.

Remember if you lie about your ambitions, you will work with the same people in the future.

This interview question is not tricky at all. If an interviewer asks this question, they are interested in knowing if they can count on you as a leader or as an expert in a couple of years, provided that you still work together.

The Role of Your Questions During the Interview

Ask questions during the interview whenever you get the chance. When researching the company, come up with a couple of questions you are interested in.

It is your chance to find out more information on your employer, your team, your technological stack, the development process, company culture, company policies, leisure activities outside work, and a lot of other things.

You can create an amazing interview experience for your interviewers by commenting on their answers and asking the next question only after the comment.

If something is not clear, it is also possible to ask questions related to an interview question. This is how you change an interview to a friendly conversation.

Remember you are equal parties, and you should get a chance to interview your interviewers just as much as they interview you. You also have to decide on whether you want to work with your employer; it's not just about them.

The more relevant your questions are, the more certain you appear on what you are looking for. This means that you *qualify* your interviewers just as much as they qualify you.

There Is More at Stake During the First Interview Than Just Passing It

The first interview is not just about a binary decision on whether the company will continue with you or not. The best possible outcome for you is that you convert your interviewers to fans. If your interviewers like you, they will help you during the technical questions if you get stuck.

A great first impression may even lead to a second bid from the company's end after you receive an offer from them. As a candidate, I have been in a position twice when I asked a day to consider an offer and I received a higher bid in my inbox without me asking for it. In one of the cases, a company sent me a contract draft where they boosted my potential salary even more.

Nail the Technical Interview and Homework Assignments

Goals of this section:

- Find ways to prepare for the technical interview.

- Find out about techniques to increase your chances even if you don't know how to solve a task.

- Learn how to pass the technical review of your homework assignments.

You will now find out more about how to approach technical interviewing. You will learn how to show your thought process and how to make your interviewers your allies. We will also discuss how to solve homework assignments and what you need to pay attention to when discussing your assignment in an interview situation.

Don't Confuse Yourself

Technical interviews are not as hard as they look like. Your interviewers are also humans as well as you are. It makes no sense on your end to be afraid of situations.

Coding interviews remind me of my university exams. Back in 2001, I had to take a linear algebra and graph theory exam. This meant to me that I had to remember all theorems and proofs to be able to pass the course and solve problems in front of an examiner that made it evident that I understood the material, not just learned it.

Learning facts and memorizing data were not my strengths back then. I didn't know about the book *Moonwalking with Einstein*[12] back then. Had I known some cool memorization techniques, I would have aced most of these exams.

Even though I studied a lot, I was afraid of topics like the theory of bilinear functions, or the proof of the statement that any map can be colored using five different colors assuming that no adjacent countries have the same color. To be exact, four colors are sufficient for this purpose, but fortunately for me, we didn't have to learn the proof.[13]

The examiners were a bunch of PhD students and some professors. There was one specific professor everyone was afraid of. Both in the dormitory and in our mailing list, information spread that this specific examiner looked at his victims with scorn whenever they were not near his level.

Ten of us started the exam in a spacious auditorium. As I opened my topic, to my horror, it was the coloring theorem. Great, I thought, and I started constructing the proof based on the vague memories I had. I started connecting the dots and constructed some parts of the proof. However, there was a point where I got stuck.

As I had no chance of recalling the missing pieces, I made a decision that I would raise my hand as soon as an easygoing examiner approaches me. I recognized a lab assistant, who always had good will and never went into details. As I raised my hand, the guy walked past me.

The dreaded lecturer noticed me and started approaching me. He greeted me with a warm smile and started studying my notes. After a minute, he asked me to start explaining it. The proof started smoothly. My thoughts were already a bit ahead, as I knew he would ask me to elaborate on the missing puzzle pieces.

I chose to admit that I got stuck. I continued by enumerating two ways of how we could continue deriving the solution.

"Let's continue with the second thought. You are looking for a contradiction, but the structure is a bit too complex to check. Think about two separate cases," my professor said.

"Cases that would both end up becoming contradictions," I added.

"Exactly. How would you restrict the structure of the graph to make your life easier?"

I started experimenting, while my examiner gave me nonverbal signs of confirmation when I found the right structure. I got to the contradiction alone. It took us some more time to go through the other structure, but eventually, the proof was done.

---

[12] Joshua Foer, *Moonwalking with Einstein: The Art and Science of Remembering Everything* (Penguin Books, 2011).
[13] www.ams.org/notices/200811/tx081101382p.pdf

My examiner started asking me random questions. As the questions got harder and harder, I started recalling the horror stories I had heard. I also noticed that my examiner became more and more worried.

After a short silence, my examiner spoke up: "I am really sorry…but I cannot give you an A, only a B."

I got the second best grade. This was an eye-opening experience for me. What exactly happened here?

Opposed to another story in the section "Your SMART Learning Plan" in Chapter 4, this time, I was the hero of the story despite confusing myself with stories that were distorted by people who justified their failure with an external cause. This external cause only existed in their imagination. The examiner was more than fair with me, and he managed to give me the grade that mapped my knowledge perfectly. He even helped me and did everything he possibly could to find out what I knew instead of what I didn't know. Most people have a tendency of demonizing their interviewers. Be objective, and focus on what matters.

Moreover, notice that soft skills allow you to form a connection with your interviewer. The key to successful cooperation was twofold: I did not even try faking knowledge that was not there. I admitted my weaknesses as if they were natural. I also expressed my interest in continuing under the supervision of my examiner. Most people give up on the spot when they face a challenge. Your strategy should be to win the support of your interviewer.

Your professional code of conduct will back your efforts up. I can tell you from my own experience as an interviewer that I always help people who do their best in expressing their thoughts and trying to show me their thought process.

You may be wondering why I chose to admit that I got stuck. My professor has been in this field for more than 10 years. He has seen thousands of students. Most students try to fake confidence, and he knows it really well. He must have developed several ways to find out the truth.

Note It is not worth the effort to make a conscious effort to impress people. If they are not worthy of you impressing them, you waste effort. If they are worthy of you impressing them, you can safely assume that they can see through all effort of you trying to impress them.

Preparation for the Technical Interview

Recall the peak performance psychology concepts of *training mindset* and *trusting mindset*. During your preparation, you are in training mindset. Once the interview starts, you have to trust your abilities. Don't stress yourself out by trying to learn something new during the last minute. Just do your best.

Some employers will not like me for this statement, but from your perspective, it is all right to gather some interviewing experience even if you don't want a job. If you have stage fright, try to do everything you can to put yourself in tough spots, so that you will be able to deliver once it really matters.

Interviews are often about generic discussions. You can expect any of these topics pop up depending on your specialization:

- Development methodologies and processes
- Technological stacks
- Git, version control
- Architectural and detailed design
- Design patterns
- Functional and object oriented programming
- REST, GraphQL
- CICD pipelines
- SOLID principles
- Database concepts (including NoSQL)
- Composition vs. inheritance
- Automated testing
- Common frameworks and libraries
- TDD, ATDD, BDD
- Cloud technologies
- CAP theorem
- Docker, Kubernetes
- Project management
- Stakeholder management
- Product management

The list is by far not complete, and your specialization may require you to study topics others don't need.

There are a lot of books and courses on practicing for coding interviews. This includes bestsellers such as *Cracking the Coding Interview.*[14]

If you are a JavaScript developer, I can also recommend my book *ES6 in Practice.*[15] Beyond the usual questions, I have also added 20 coding interview questions and answers. You will get a chance to practice JavaScript interview questions and formulate thought processes.

Regardless of the programming language of your choice, I can highly recommend a couple of web sites to you, where you can fine-tune your coding skills free of charge.

Codility[16] is my number one recommendation. Codility guides you through the fundamentals of algorithms and summarizes the key principles you need to know as a software developer. It starts nice and easy with iterations and arrays and provides you with a comfortable learning curve, culminating in dreaded topics like greedy algorithms and dynamic programming. Codility is a must-have for you if you are targeting prestigious jobs. By committing to one challenge a day, you will drastically improve your coding skills within months.

HackerRank[17] is number two on my list. The interface of Codility is simply nicer, and given the lack of quality control in submitting exercises, some tasks tend to be harder to read.

If you want to get a job with the big four, you might want to consider Project Euler.[18] Be careful though, as the tasks will get very mathematical, and you might spend too much time on tasks you won't ever need.

Coding interviews are not only about math and algorithms. LeetCode,[19] for instance, supports database and shell-related questions as well.

My personal opinion is that technical interviews are often flawed. Some interviews don't require you to write any code. I have attended multiple interviews, where the whole experience was on a superficial level. If I claimed I knew the right technologies and my soft skills were adequate, I got hired. This worries me as a candidate, because the hiring process indicates the skill level of your future team.

---

[14] Gayle Laakmann McDowell, *Cracking the Coding Interview: 189 Programming Questions and Solutions*, sixth edition (CareerCup, 2019).

[15] Zsolt Nagy, *ES6 in Practice: The Complete Developer's Guide* (Leanpub, 2018). You can find it on http://zsoltnagy.eu

[16] www.codility.com

[17] www.hackerrank.com

[18] https://projecteuler.net/

[19] https://leetcode.com

The other extreme where programming interviews go wrong is a bias toward algorithms. I get that software developers should be creative. Hard algorithmic puzzles don't necessarily test the algorithmic skills of developers. At the top of the hierarchy, you rather test whether the candidate solved a similar problem before.

Think about it. Dynamic programming was not invented in 20–30 minutes. This is the exact time limit you have for a dynamic programming exercise. If you have no clue about what dynamic programming is, good luck meeting the complexity requirements of a coding exercise.

A deviation of the coding puzzles is when interviewers test whether a software developer knows the ins and outs of edge cases of a programming language.

We can debate whether coding interviews are fair or not, but this debate won't get you hired. We will therefore focus on what matters from the perspective of our career and accept the world as it is. Research the interview experience you expect, and practice accordingly.

If you want more practice, look for career centers or agencies that offer practice interviews. At the time of writing, I offer a free consultation on my web site.[20]

Coding Interviews

The single most important thing in coding interviews is to show your thought process and cooperate with your interviewer whenever you can. You can recall my university exam story. Interviewers want to help you as well as my professor did.

Put yourself in the shoes of your interviewer. Before you, he might have rejected ten developers in a row. I have interviewed several hundred candidates, and I rejected around 70% of them during the first interview.

Many software developers cannot solve simple tasks. Whenever I see that someone is in command of writing code, I get energized and do everything to make the interview experience of the candidate easier by correcting their typos and helping them with their research.

If you express your thought process before solving a task, you get access to valuable feedback. Involve your interviewer in the process of constructing the solution.

Once I was interviewed for a frontend developer position. MongoDB was new back then. The CEO asked me a question: "We are using both MySQL and MongoDB. When would you use MySQL, and when would you use MongoDB?"

---

[20] http://devcareermastery.com/free-interview/. The author reserves the right to discontinue this offer at any time.

I started expressing my thoughts slowly: "Let me think about it. I experimented with MongoDB once, so I can add what I already know. When I used MongoDB, I could store JavaScript objects in the database. Back in the days, I studied a bit about object oriented databases, about the differences and the use cases … SQL and tables are simple and straightforward."

My interviewer interrupted me. He said, "Think about the structure of the result of a select statement, and compare it to JSON data."

I immediately replied, "In a relational database, data are well structured, and they are optimized for retrieval. While in an object oriented database, elements of a result set are not structurally restricted. I can imagine that you have to integrate data sources of different kinds. It is a lot more convenient to handle objects than to patch the database schema of a relational database to support these types of objects and associations, especially if the data structure changes often."

The key element of the interview was that the interviewer helped me. He not only confirmed that my initial thoughts were somewhat useful, but he also gave me a hint.

Sometimes showing your thought process and finding out the answer to a question is more convincing than just knowing the answer by heart. Once you draw the right conclusions, your interviewer will most likely be impressed.

Not all interviewers will help you though. Sometimes they even leave the room as you do the coding exercise. If your interviewer stays with you while you code, make sure you always announce what you are thinking about.

Always verify your solution, and polish your debugging techniques. Know when you log and when you use breakpoints. Know what values you look for and what you expect as a result. Always test one assumption at a time. As soon as you are done with one assumption, clean up your code immediately.

During a coding task, it takes a lot of courage to admit that you are on the wrong track. If you can see that your code is growing and growing and you start losing track of what is going on, sometimes a hard reset helps you. We often get an easier idea after trying out a harder, less straightforward path. Surprisingly, only a fraction of developers get rid of their bloated code in favor of a better idea. Patching code that doesn't work is hard. You will often increase your chances by restarting the task using a different thought process.

Coding skills are not enough in a technical interview. You also have to master expressing your thought process. Communication is not always easy. If you want to get better, you can do any of the following tasks:

- Record yourself explaining an algorithm, and listen to it afterward.

- Tell a software developer friend of yours about an algorithm.

- Formulate a process around an activity you like doing. Express it as if it was an IT process.

- Write proper documentation and blog posts.

- Practice interviews even if you don't plan getting hired.

Don't forget to ask as many useful questions as possible. Qualify your interviewers, and make sure you find out everything you need for making an educated decision. Recall the case study from the beginning of this chapter, and remember the importance of your questions. In a technical interview, you can find out a lot of things about your future employer.

Lastly, the more interviews you attend, the more likely it is that you stumble upon unfair questions. As interviewers are also humans, it is inevitable that some of them will come up with nonsense. Whenever I observe the lack of competence of my interviewers, I know that the company is not for me; therefore, I reject them on the spot at the end of the interview. Fortunately, this rarely happens.

In case you get rejected, you might have heard that it is possible to ask for feedback on how the interview went. Many companies don't give you feedback, as they either don't have time to write one or they are not comfortable with it.

You can still maximize your chances of getting feedback by resending your feedback request a couple of days after your first email.

If a technical interview ends with success, you will get one of the following three results:

1. An appointment for another technical interview

2. An offer

3. A homework assignment

We will now continue with homework tasks.

Homework Assignments

Homework assignments may vary a lot. I have heard of companies who expect you to write a poker bot with artificial intelligence. Other no-name startup companies send you a hard task. Once you submit your solution, they reject you and utilize your free work. It is very easy to recognize these fraudulent interviewers, and fortunately, there are not too many of them.

Most of the time, your homework assignment will be reasonable. In order to ace the programming assignments, follow these rules.

Read and interpret the whole exercise in depth. Make sure you understand every little detail of the task. If something is unclear, ask the company. *Some companies underspecify their tasks on purpose, just to test how proactive you are.*

Solve exactly what is requested, nothing more, nothing less. Never invent features that are not requested. They will rarely be appreciated; they just create noise on top of your solution. If you fail to complete some of the tasks, you will be at a disadvantage against other developers who complete all features. You can still get hired, but your solution should then stand out in another way.

Write maintainable code. Use your favorite linter, comment your code, and make it consistent and readable. Design your solution with maintainability in mind. Show maturity. Avoid hacks. Excellent books have been written on this topic:

- Robert C. Martin: *The Clean Coder: A Code of Conduct for Professional Programmers* (Robert C. Martin Series)

- Steve McConnell: *Code Complete: A Practical Handbook of Software Construction, Second Edition*

Name all your sources. You don't have to reinvent the wheel. You can use boilerplates or other repositories written by others as long as they bring you forward. However, you have to justify your choice and name your sources. Never use code of anyone without giving them credit.

Document your solution. About 30% of the solutions I received never came with installation instructions. Sometimes I could not launch the application. Other times, I received tasks with global package dependencies that were not installed on my computer. I had to manually determine which packages to install so that I could run and evaluate the application.

Write bug-free code. Go the extra mile and debug your solution. Your interviewers know the ins and outs of the challenge they give you. This means that if you ship them buggy code, they will find your bugs. If you send them the message that you are not capable of writing a small application without bugs, they will start doubting your capabilities of writing good enough code.

Tool up. Use frameworks, libraries, and boilerplates. If your solution tells the story that you keep reinventing the wheel over and over again, you will be regarded as a junior developer, which won't help you during negotiation.

Automated testing. If you add test coverage to your homework, it is often a big plus.

Study your solution. If your homework assignment is accepted, be prepared to receive questions on it. For this reason, make sure you can run and explain your code before starting the second interview.

Negotiate Effectively

Goals of this section:

- Learn the psychology of salary negotiation.
- Equip yourself with negotiation techniques to maximize your earning ability.
- Salary negotiation myth busting.
- Find out how you can practice negotiation.

This section is about effective salary negotiation. We will demystify the popular belief of always countering twice. You will read about strategies backed by common sense.

Let's start with the psychological background of salary negotiations. You will find out who your allies are, and you may draw surprising conclusions. You will then learn about myths generated by the career self-help industry.

When career coaches spread some myths, they might not realize that their clients were sometimes successful in spite of and not because of their strategy. Remember software developers are often involved in the recruitment process. Many of them have heard of mainstream advice.

Once you can differentiate between common-sense advice and shiny myths, it is time for you to learn some negotiation techniques. These techniques are in perfect alignment with the contents of the rest of the book.

We will conclude this section with some advice on how to practice negotiation.

The Psychology of Salary Negotiation

Many software developers negotiate their salary once every 2 years. This is hardly enough to gain a competitive edge. Therefore, many developers are very bad at negotiation.

Negotiation usually happens during the end of the interview process. You may already suspect that your chances are mostly determined by your performance during the interviews. However, there is some money to be made during the final stages.

From the point of view of the applicant, your interest is to avoid committing to a range early and avoid looking needy. Don't reveal that you need a job badly, as it may easily cost you money. Your interest is to excel during all stages of the interview and impress your interviewers with your personality and skill set. If you get the fundamentals right, you have no reason to worry about the outcome.

■ **Note** Companies are interested in finding out if their budget for their role is compatible with the salary expectations of the candidate. The earlier they find this out, the less they have to invest in candidates they can't realistically recruit. Candidates are interested in delaying revealing their expectations until they manage to hook their interviewers with something that makes them unique.

Prices are shaped by demand and supply. The more unique and valuable a candidate is, the lower the supply from what the candidate can provide. The better the candidate's skills match the company's needs, the higher the demand. Increased demand and decreased supply drive up the price.

Similarly to stock trading, poker, or any game where money is involved, two twin forces shape our actions: the *fear* of losing the deal and the *desire* of gaining more resources. Tony Robbins calls this the pain-pleasure principle and adds that for most people, pain is a bigger motivator than pleasure.

As a consequence of being afraid, candidates typically make some of the following mistakes:

- They reveal a lower sum or range than their true value or desire.

- They tell their interviewers that they are looking for a market value compensation and will accept the offer of the company.

- They never counter an offer even if they got lowballed.

- They don't impress their interviewers well enough, because they are afraid of saying something wrong.

- They don't form personal connections, as they appear to be afraid of the interview situation, making both parties uncomfortable.

Some software developers learn how to *hide* fear by following advice of career gurus, faking confidence with some techniques. The good news is that this is typically a more successful strategy as being fully passive. The bad news is that you may undermine your relationship with your future colleagues. As you are interviewed by intelligent people, many of them will see through your attempts. These attempts are as follows:

- Bringing up salary and other benefits early

- Continuously bragging about yourself (see implicit vs. narrative influence)

- Countering twice, even if you had committed to a lower figure previously, even if you got a salary above your market value

- Applying ruthless and exploitative negotiation techniques without considering the big picture

Note Many companies tend to reject developers who seem to be high maintenance. If a development manager has been burned by having dealt with an employee having cluster B personality disorder traits such as narcissistic personality disorder, Machiavellianism, or psychopathy, they will screen for low-maintenance employees, and some of the fake techniques will work against you. Techniques that look fake are not integrated in your personality. These techniques make you appear to game the system. Gaming the system is a Machiavellian approach.

Your interviewers tend to experience similar feelings as you do. They also fear of losing great candidates, and they are motivated in getting a great deal. The intensity of these feelings is different though. There is almost always a lot less at stake for your interviewers during the beginning of the interview process. As you appear more and more valuable, you reach a point where your unique value will be more real to your employer.

Note Recruiters earn 15–20% commission on your yearly salary. If you make $100,000 a year, this is $15,000–$20,000. In average, the cost of hiring a competent developer is in the five-figure range. This means once you get an offer, chances are that this five-figure opportunity cost will be associated with your hiring decision. This is why you often have negotiation leverage, simply because companies don't want to invest another five figures to find another applicant who is a good match for them.

You are in a comfortable situation when your interviewers need you. In other cases, your interviewers tend to sit back and subjectively evaluate your skills and salary expectations.

There are two cases when the hands of your negotiation partner are tied:

- Some companies have fixed starting salaries. You may get away with negotiating some small benefits or flexibility, but your salary is fixed.

- Your interviewers may be bound by budget constraints that don't allow them to fulfill your salary expectations. If you are exceptional, they might ask for extra approval, but miracles rarely happen.

Once I interviewed for a position with a startup. By the end of the second interview, they really liked me and wanted me to get on board. Unfortunately, their budget was capped at $150,000, and my minimum expectations were $200,000. The deal was not possible, even though they were already planning how they would utilize my experience on multiple levels.

In other cases, there is some money to be made. My personal experience is that your negotiation partners are often on your side once they know that you are an above-average developer. Many interviewers realize win-win situations, and they understand that everyone wins if they let you win. In a situation like this, aggressive negotiation makes little sense, as your negotiation partner is on your side. Being assertive is superior to being aggressive.

It is very easy to blame your negotiation skills for getting lowballed. In practice, these are the top two reasons why you get a significantly lower offer than your expectations:

1. You didn't do a good job during the interviews. As you didn't create a lasting impression, you are not regarded as a hot commodity.

2. Your employers want to lowball you on purpose regardless of your negotiation skills. They are takers,[21] and they want to take advantage of weak negotiators instead of paying more for premium services.

In case 1, the mistake was made before negotiation, and you have to improve your interview performance. In case 2, the only mistake you made is that you had not interviewed your interviewers well enough to determine that you were better off ending the interview process with them. You still gained valuable experience in exchange for the time invested on your end.

Realize that your negotiation partners are on your side more often than you think. The better professional you are, the less you have to focus on negotiation. Good companies cannot afford to underpay knowledge workers. As a consequence, win-win situations are widely available.

---

[21] More on this topic: Adam Grant, *Give and Take: Why Helping Others Drives Our Success* (Orion Publishing Group, 2014).

Note   Hiring managers are employees managing a budget. The first rule of managing a budget is that a budget is either loosely tied to the KPIs (Key Performance Indicators) influencing the salary of the hiring manager or it is not tied to these KPIs at all. The second rule of managing a budget is that budget has to be spent or it is lost by the end of the financial year. Although constraints may prevent hiring managers from taking extreme decisions, when they find a very good candidate, they support enabling the best possible treatment for them.

Intellectual Fog Around Negotiation Advice

Not too long ago, I went back home to spend Christmas with my family. I also had some alone time in the room where I grew up. I stumbled upon my university notes and found a file about my first ever IT career workshop. The most prominently highlighted sentence of my notes was "The first person to reveal a number loses the negotiation."

A couple of months after taking this course, I was asked to reveal my salary expectations. I executed the routine I had learned and practiced thanks to the career workshop. I can still remember the smile of an experienced CEO sitting opposite me. He said, "Oh yes, you believe that the first person to name a number loses."

Note   Negotiation advice uses the concept of **information asymmetry** to imply that you lose by revealing your expectations. This is because companies have access to a lot of data, while you don't. In the age of Google, Glassdoor, and PayScale, this advice seems outdated to me. However, your employer still knows more about their constraints than you do. Negotiation advice also forgets to consider that the higher your skill level, the less information asymmetry applies to you as a disadvantage. As this book is written to help solid, good professionals, chances are you end up in a situation where multiple companies bid for your services and you are in a position to override budget constraints.

Does the first person to name a number really always leave money on the table?

Sometimes yes. If you name the wrong number, some people will take advantage of your lack of research. This is the exact reason why I asked you to do your research before negotiation. In the 21st century, enough information is available to judge how much you are worth for a given company.

When in doubt, it makes sense to attend multiple interviews and ask for a slightly higher starting salary than the amount your research indicates. You can learn from rejections, and who knows, maybe you hit the jackpot.

For instance, as a fresh graduate, I had an interview experience with a less experienced HR manager. I didn't really want to work with them. I wasn't sure, but I had the impression that their budget was quite low for fresh graduates. When she asked me to reveal my salary expectations, I named a high number. The HR manager couldn't hide her surprise. She started saying that maybe with the food ticket system, training budgets, and additional benefits, they could get close to that number, but it was unlikely that they could pay this amount. Then I told her that it was very important for me considering that this job required a significant specialization on my end. Afterward, we terminated the interview, and she asked me to get back to her once I considered a lower starting salary.

What happened in this interview?

I named a high number. My negotiation partner revealed a lot of valuable information on their budget. Then they tried to reach an agreement with me. Obviously, their budget constraints didn't make it possible to close the deal.

I walked away with valuable information, and had the company been less restricted about their budget, I could have ended up working for them.

When I started my freelancing career, I was shocked at the low rates offered by other skilled freelancers in popular freelancing sites. I was even more shocked to see clients post projects where they expect software developers to put in a lot of work in exchange for a two-digit payout, and many developers actually applied so that they gather the feedback that gets them started. Once I had a client who was impressed by my relevant experience, and he asked for my rates. I named $120/hour, and the client accepted it even though he could have selected hundreds of developers for the same job in the $30–$50 range. I really didn't mind losing the deal; in fact, I expected not to get the gig. To my surprise, the client unconditionally accepted my terms and utilized my services for about half a year, for 20–40 hours a month. With a stronger personal brand, it is possible to charge $200, $500, or even $1,000 per hour.

Even though these rates seem high, it is possible to go even higher by applying my favorite strategy: charging per project and guaranteeing a result. This way, the freelancer takes the risk and reaps the rewards in case of early completion. The client outsources the completion of a hard task, and they don't mind how you complete it. In this case, I come up with the terms, the guarantee, and I determine how much I charge.

In the case of the $120/hour example, I could have received an initial offer of $50/hour. If I then say $120, I will have to be a very good negotiator not to lower my rates. If you charge on a project basis, you also need good experience with negotiation.

Note If you expect your negotiation partner to lowball you either on purpose or due to lack of information on you, *not naming your expectations first may hurt your chances* especially if you are not an experienced negotiator.

Most people fear that naming a slightly higher number than the budgeted range would hurt them during negotiation. This is sometimes true; however, the more you have to offer, the stronger your position becomes during negotiation.

Recall the principles of professionalism and taking responsibility. Regardless of your experience level, you are responsible for researching your salary expectation. Then you are responsible for communicating it to your potential clients.

People who avoid taking responsibility ask for the company to give them a reasonable offer. This offer may be reasonable at best. You will rarely receive an extraordinary offer. All you reach by hiding your expectations is that the company will reveal a figure you can find on the Internet.

According to another negotiation myth, you always have to counter, often multiple times. The more you are in this business, the more you know when your client or potential employer reaches the limits of their options. If you do an extremely good job during the interview, two things happen:

1. You start gaining more by naming your price yourself.

2. Your employer may admit that they give you the absolute maximum they are authorized to give you.

While there is some money to be made even when you are presented with facts that look nonnegotiable, in some cases, there is just no more room for further gains.

Another myth is about always hiding your expectations until the very end. In a few cases, this will eliminate you from the process. In other cases, you may end up hiding your expectations from people who are on your side. For instance, if you work with an external recruiter, their interest is to make more money. If you target a higher salary and they get 20% commission on your salary, it is mutual interest for both of you to maximize your salary. This is a source of valuable information where the recruiter may give you insider information.

In some cases, no excuses save you from entering your expectations. The easiest example is a form where the minimum value you can enter is the visa minimum you can earn, which is a reasonable salary in many countries. The maximum is uncapped. You cannot enter zero, you cannot enter a negative number, and a system filters out candidates automatically who enter more than the top of the budget. Contacting the company with guerilla tactics does not pay off either, because everyone is obliged to execute the same process.

Note If a company insists in you entering your expectations, you have the right to change your mind later, especially if the interviewing process reveals some information that makes your position more responsible.

This argument concludes the myth-busting section. These arguments do not represent the universal truth:

- The first person naming a number loses.

- Countering is always worth it when you receive an offer.

- You can delay revealing your expectations until the end of the interviewing process.

I am not saying that these guidelines never work. But I am implying that these guidelines do not always work in your favor. Sometimes you have to apply your own judgment to optimize what works in your specific situation. Any universal rule counts as intellectual fog that is intended to prevent you from seeing reality.

Negotiation Guidelines

Although it is not always the case, it is often beneficial for you to delay negotiation to the very end of the interview process. If your interviewers ask about your expectations during the first interview, before they know anything about you, you can turn the situation around. You can tell them that you know hardly anything about the company, you have not met many of your colleagues, and therefore, you don't have enough information on giving them a number. You can then recommend that you get to know each other a bit better during the interviews, and then, it will be a lot easier for both of you to close a mutually beneficial deal. A shorter version of the same message is that you are still researching the situation. This is a valid reason especially if you move to a new state, city, country, role, or industry.

Sometimes your negotiation partner will insist in getting your expectations. Your interest is then to use your own thirst for information to blur your expectations to a range, depending on many factors. The idea behind giving a range is that you make the end result depend on many factors that will be revealed later during the interview process. However, the stronger your personal brand, the higher the likelihood that a well-researched high number pays off for you in many cases. The closer you are to the end of the interviewing process, the more confident you can be to name a number and not a range.

There is an option to form an alliance with your negotiation partner saying that you know that her job is to make sure your claims are not unrealistically high. Therefore, if she tells you their budgeted range, you can tell her if it makes sense on your end to continue. Yes, we got back to the famous "first person to reveal a number loses" rule; however, in this special case, you just negated an unfair disadvantage. Once you get an answer, you should tell her that you are confident that you will reach a conclusion. You don't even have to reveal a number, and you don't have to imply that your confidence comes from expecting a salary within the range or outside it. Make sure you don't say that your expectations perfectly fit in the range, as it limits your upside potential later.

The third option saves you some time. You can reveal a more accurate clue about your expectations. If your negotiation partner becomes worried or expresses a mismatch in expectations, you can tell her that you know these figures are a bit high, but your job during the interviews is to prove that your contribution is more than worth this figure. Furthermore, you believe in lasting relationships; and therefore, it makes little sense on your end to commit to a job that pays below market rates.

Some companies ask you to reveal your expectations before the first interview in writing. When this happens, it is all right on your end to write "Negotiable." This may or may not work. Worst case, the company will get back to you. In that case, you will have to enter a legitimate number, not zero, not 999999, but your real expectations. You maximize your chances by making it clear that the figure you entered is only preliminary and not binding, as you need to go through the interviews to collect sufficient information on your expectations. Remember you have the right to change your mind, especially if you encounter new information.

Some companies tend to ask you to reveal your current salary. It is absolutely fine to make it clear that this information is confidential. You don't have to explain yourself. Your current salary can be used against you, especially if it is lower than the amount you are targeting. In many countries, this counts as an illegal question, especially if your contract states confidentiality.

Don't burn bridges behind you. Even if negotiation does not work out in the way you wanted to, never insult a potential employer. They also have connections, and the world is too small to deal with the potential consequences of lack of professionalism on your end. Whenever an agreement is not reached, simply thank your negotiation partner for their efforts, and move on.

Be honest. Don't hide behind negotiation techniques that you cannot present with integrity. If you are a good professional, your employer will have a lot more to lose than you by not hiring you or underpaying and losing you. Personal integrity has a lot of benefits.

For instance, a friend of mine got a job in London. After he got the offer, he asked for a 20% higher number based on his own research. He was confident that he deserved this amount. He even talked to the representatives of the company once more showing evidence that his claims are not false. Eventually, he got the amount he requested.

He found out a year later that his salary was the highest in his team. He also found out that HR had to get special approval for hiring him. However, his performance was so good during the interviews that HR thought it was worth for them to get extra approvals.

Practicing Negotiation

Always stretch your limits. The best way to stretch is to gather hands-on interview experience. You can negotiate with confidence in situations when stakes are low. If you have a hard time putting theory into practice, send out a couple of applications, and start interviewing. Just make sure you don't select the company of your dreams for practice.

Negotiate in everyday situations. When you go to the market or order products, try to negotiate small discounts. You will get better with practice.

There is one exercise that helps you a lot with developing your confidence: record and rewatch yourself as you negotiate in situations, where recording the negotiation is legal. First, it will be very uncomfortable. If you do it 10–20 times, you will get more familiar with your own nonverbal communication. You will also be more comfortable with yourself than ever before.

Checklist for Targeting a New Job with Experience

Let's assume that your decision is final and you are thinking about leaving your current company for a good reason.

- ✓ **Design your career path** (Chapter 3).
- ✓ **Determine the next step you would like to take** (Chapter 3).
- ✓ **Evaluate if your current employer can help you take the next step.** Before starting the application process, investigate what it takes for you to get promoted (Chapters 5–6).
- ✓ **Gather information on positions matching your expectations**. Conclude what you need to learn in order to get ahead (Chapter 7).

✓ **Come up with a learning plan.** Create meaningful projects, preferably open source. Blog about the things you learned. Go to relevant meetups (Chapter 4).

✓ **Update your social media presence** and your basic resume based on your personal brand (Chapter 4).

✓ **Organize your thoughts; shift your mindset and your attitude.** This helps you deliver peak performance during the interviews. Don't forget your current role either; treat them as your clients (Chapter 2).

✓ **Execute your to-do list for each position** (see in the following) to get offers.

✓ **Resign once you accept an offer.** Make sure you only accept one offer. Resigning early only makes sense if you know for certain that you will easily find a new job.

For each position:

✓ **Research the company,** and find out if their offer is promising to you. Find testimonials or insider information (Chapter 7).

✓ **Tailor your application package based on the needs of each company you apply for one by one.** This includes your resume, motivation letter, and list of relevant projects. Use your personal brand (Chapters 4 and 7).

✓ **Prepare for your interviews in advance.** Work on your self-esteem and self-image, examine your fears and limiting beliefs, and make sure your communication is polished (Chapters 2, 5, 7).

✓ **Practice and prepare** for the nontechnical and technical interviews (Chapter 7).

✓ **Negotiate** whenever you are asked about your salary (Chapter 7).

✓ **Find out as much as possible about the company.** An interview is not only about you. You should also determine if you would like to work with a company. Reject them as soon as you find out a company is not meant to provide a good atmosphere for you (Chapter 7).

✓ **Gather feedback** using your communication skills (Chapter 5).

✓ **Evaluate offers.** The more items you have completed with quality work, the more likely it is that you will be able to choose from between multiple offers.

✓ **Determine if it makes sense to negotiate** before accepting an offer. If you are planning to reject an offer anyway, you have nothing to lose (Chapter 7).

Checklist for Targeting a New Job Without Experience

When you don't have relevant experience, your situation is slightly different. You are either a fixed college or university student with no experience, or you are planning to change careers.

I will assume that you don't have a job at the moment. If you already have a job, make sure you negotiate your notice period with your employer, and resign once you get an offer.

You will most likely start as a generalist junior developer. Designing your career path always helps, but it is not essential for your first job. You may gain some bonus points by being able to answer how you see yourself in 5 years, but let's be honest. With minimal relevant experience, this is not the time to worry about such questions.

✓ **Design your career path** (Chapter 3). Think in terms of at least 3 years, preferably 5.

✓ **Come up with a learning plan, and build your network.** Build your portfolio. Create meaningful projects, preferably open source. Blog about the things you learned. Go to relevant meetups (Chapter 4). Substitute demonstrated working experience with demonstrated skills.

✓ **For each hour you study, allocate 2 to 4 hours of practice.** Learn by doing. Build your portfolio. An uploaded open source project you talk about is worth a lot more than any certifications.

✓ **Social media.** Relevant studies and experience are valuable on your LinkedIn profile. Testimonials on your reliability and soft skills are often positive. Meaningful tweets are appreciated. Social media will not be a big factor in getting a job, but you have to start somewhere (Chapter 4).

✓ **Train your soft skills, mindset, and attitude** (Chapters 2 and 5). These skills will come handy during the interviews.

For each position:

✓ **Research the company, focus on gaining career capital quickly,** and find out if their offer is promising to you. Find testimonials or insider information. You are not likely to earn a fortune with your first job; therefore, either look for valuable skills to study or a career path inside the company (Chapters 3 and 7).

✓ **Tailor your application package based on the needs of each company you apply for one by one.** A motivation letter demonstrating high emotional intelligence and the completion of some tutorials and side projects demonstrating expertise will get you ahead of 90% of the applicants without experience (Chapters 4 and 7).

✓ **Practice and prepare** for the nontechnical and technical interviews (Chapter 7).

✓ **Negotiate** whenever you are asked about your salary (Chapter 7). However, be ready to sacrifice your earning ability as a junior developer, in exchange for gaining career capital. Make sure you show that you are a person ready to cooperate with the company. Be assertive, but stay realistic. If the company is not ready to pay you the salary you hoped for, make sure you discuss terms under which it will be a win-win situation for both the company and you to work under the terms you imagined.

✓ **Find out as much as possible about the company.** An interview is not only about you. You should also determine if you would like to work with a company. Reject them as soon as you find out a company is not the right fit for you in terms of atmosphere or career capital (Chapter 7).

✓ **Gather feedback** using your communication skills (Chapter 5).

✓ **Evaluate offers.** The more items you have completed with quality work, the more likely it is that you will be able to choose between multiple offers.

✓ **Determine if it makes sense to negotiate** before accepting an offer. If you are planning to reject an offer anyway, you have nothing to lose (Chapter 7).

Summary

Regardless of whether you have experience or not, you need to go through the same obstacles when interviewing for a position. You can expect to talk to HR, businesspeople, and engineers. You need to connect with all parties.

Your best strategy is to go the extra mile, show genuine interest, and respect everyone you meet even when the interview appears to be over.

When you submit your resume, make sure you tailor it to the position you are applying for. In the behavioral interview and in the tech interview, your best strategy is often to be transparent with your thought process. This way, your interviewers will be able to help you.

While negotiation is an important part of your application process, the most rewards are earned during the application process. Once you make your life as easy as possible, you will have an advantage when it comes to negotiating a higher salary for yourself.

Don't forget the checklist that applies to you at the end of the section to prepare for your next interview.

Your Future Is in Your Hands

Software Development and Education in the 21st Century

Work environments continuously change, and this change benefits talented developers.

In the last century, most people had to get used to 9–5 jobs, sometimes with overtime. Many jobs let you choose the time of the day when you work. Many jobs put heavy emphasis on work-life balance.

While some developers work in offices, more and more people choose to work remotely. Remote work is made possible with continuously improving collaboration software. This trend will continue with Virtual Reality products in the future.

Emphasis is shifted from traditional employment to entrepreneurship. Entrepreneurs are rewarded for taking risks and assuming full responsibility. The same holds for employees who assume more responsibility than others.

Talking about results, your achievements will matter more than time spent in the office. Based on the principles of assuming responsibility, more and more people will be paid based on their achievements and not based on an hourly rate.

© Zsolt Nagy 2019
Z. Nagy, *Soft Skills to Advance Your Developer Career*,
https://doi.org/10.1007/978-1-4842-5092-1_8

Corporate structure is changing too. In the last century, your utility was proportional to the number of people you managed. This forced some great experts to become managers. Today, an expert may earn more than their team lead or manager. The more you specialize, the more you will be in demand by a smaller selection of potential employers.

When it comes to specialization, instead of "climbing the corporate ladder," chances are you will be better off creating your own ladder by taking charge of your learning plan, your specialization, and offering specialized services instead of employment time.

If this is the future you imagine yourself in, the good news is that we are laying down the foundations for this future in this book. Instead of focusing on how to squeeze out the last cent out of your current or potential employer, we are focusing on creating value and increasing your earning ability to a level that you would otherwise never be able to reach.

A Career Worth Pursuing

Your professional career lasts for around 45 years. If you waste your potential in a dead-end job, ask yourself the costs of not doing anything for 1 year, 2 years, 5 years, or even 10 years. Some people waste 20% of their careers by not paying attention to their choices.

If you are in a bad situation right now, you have received enough action items to improve your situation.

If you are in a relatively good situation, ask yourself what you will miss out on in 5 years if you stay where you are. Only a few companies give you a career path worth pursuing. Other companies are interested in keeping you exactly where you are right now: in your comfort zone and, more importantly, in the exact same position doing the exact same work as before.

Chances are that you will realize a problem about your current situation in 1 year or 2. Once you get to this state, you will likely become desperate for a change. Don't wait for this stage of your life; start planning your future right now.

Beyond employment, there are other lucrative options to consider. For instance, you can write a book, create a course, start part-time freelancing, or create a software-as-a-service product.

Autonomy and Alignment

In the 21st century, collaboration has become very important. You have to collaborate with your team to reach a common goal. You are not only a small cog in a big machinery but an important member of your team. Your opinion matters regardless of your position.

There are two orthogonal factors that characterize the success of your team: autonomy and alignment. The higher your level of autonomy, the more responsibility you can assume. Taking more responsibility is mutually beneficial both for you and for the company, as long as you assume responsibility in areas that bring the company forward.

How can we make sure that your efforts point toward the right direction? By creating alignment. Your actions and the actions of your teammates should point toward a common goal.

If you know what goals you should be working toward and you also get complete freedom to contribute to these goals in the way you want, your work will become more meaningful than before.

If you couple this experience with conscious career planning and continuous opportunities in learning and development, you will have a hard time leaving your company.

One framework that keeps people motivated are OKRs (Objectives and Key Results).

OKRs enable you to contribute to *business objectives* through influencing performance metrics referred to as *key results*. Key results are measurable, and it is fully up to your team how you reach this number.

For instance, imagine that your task is to improve the page load of your web site from an average of 1.92 seconds to 1 second or less. The path is fully up to you. You can use caching techniques. You can improve the server-side code of your API, optimize SQL statements, or use a better ORM. It also makes sense to review your client-side packages and optimize your resources.

Tasks like this are a lot more challenging than simply fixing some bugs. You will learn more, and you will eventually become a better professional.

Your own performance will be fully visible. The company may reward you, they may promote you, or you can ask for getting a raise or a promotion in exchange for a higher accomplishment. The system has benefits for everyone.

There are obviously downsides to OKRs.

First of all, it is hard to set it up. Not all company sizes benefit from it. Especially in the case of small startups, by the time you define your OKRs, you may have to firefight some issues.

It is a lot harder to hide among the pack of developers, as your contribution becomes more visible.

When the framework is implemented wrongly, there are some teams who can hack the system by consciously selecting the path of lower resistance.

Some goals are harder than others. When you do it for the first time, you may feel you have little to no influence in reaching that goal.

On the other hand, where some people see problems, others see opportunity. Regardless of whether you reach your target or not, your learning curve will be significantly faster than in another environment.

A common pattern for implementing a system like OKRs is that venture budget is allocated on it to make it work. Some systems work very well, while others only work well on paper. When OKRs don't work out well, it is often due to a clash between the vision and goals of management and the development teams. For instance, when technical debt is very high, management often doesn't understand the need and urgency to partially repay this debt before moving on. When engineering says that maintenance is needed, it is also the interest of the business to prioritize some housekeeping in case they have some long-term plans with the maintained software.

Check out the working environment of Spotify in these two videos:

- **Part 1**: https://vimeo.com/85490944
- **Part 2**: https://vimeo.com/94950270

Even though the presented structure has been improved since then, the way the role of autonomy and alignment was presented there is worth noting. These videos are not about OKRs. These ideas just make implementation of an OKR framework smoother.

In the first part, there are many interesting takeaways. Back when the video was published, Spotify worked with *autonomous squads* having end-to-end responsibility for what they build. The autonomous teams support business goals. Autonomy is not only motivating, but it is also fast, as the overhead of coordination is eliminated. Most decisions are taken in the squad.

In traditional hierarchical team structures, teams collaborate and communicate with each other. When a team is blocked by another team, prioritization of the other team may be detached from the main business goals. As a result, efficiency tends to be lower. By giving full autonomy and responsibility to squads, things move forward, as each member of the squad will be dedicated on supporting the same goal full time.

Autonomous squads should be tightly aligned. The higher the alignment, the more autonomy a company can afford to give its squads. In an environment where trust is more important than control and the negative effects of politics and fear are eliminated, people are a lot more motivated and productive in the long run. Fear may boost productivity in the short run, but it backfires in the long run. You want to work with intelligent people, who perform at their peak under good working conditions.

Obviously, having squads, tribes, chapters, and guilds may sound like an overkill for you, potentially resulting in conflicts of interest and multiple people managing you. In a low-autonomy environment, this is exactly what would happen. In a high-autonomy environment based on mutual respect between parties, more information does not mean more orders. More sources of information just mean that we can take educated, better decisions.

If you recall the Toyota Improvement Kata board at around 11 minutes of the second video, you can conclude how OKRs can be made easier. Define the current state and the desired state. Come up with the next target, and introduce metrics for measuring when you get there. Lastly, you identify target actions that move you toward the target condition. This is not the only way to do administration, but it is one way that works.

Redefining Work in the 21st Century

In Chapter 6, you have seen that your salary is determined by demand and supply. As I am writing these lines, software developers are in high demand and supply is limited. The limited supply is not about a shortage of applicants. It is rather about a shortage of qualified applicants.

When prices are shaped by growing demand, working conditions are often dictated by the participants on the supply side: software developers.

It is getting harder not to see the latest trends: the option for remote work, flexible working hours, and the shift away from the 8-hour workday.

Job boards such as https://remoteok.io/ and https://stackoverflow.com/jobs/remote-developer-jobs have emerged, where six-figure developer jobs are available, in exchange for working from the comfort of your home. Development managers get regular requests from their employees to do remote work for at least a few weeks a year when they go on a long holiday. For instance, in Germany, winters are long and cold. Some of my colleagues coming from warmer environments suffered during this cold winter, and some of them chose to spend their winter somewhere warm. Other companies around us provided the same benefits.

Back in the days, I had to be in the office by 9:30 a.m. I can still recall an uncomfortable meeting when I arrived at 9:50 a.m., because there was a medical emergency at home and I had to secure Calcium Sandoz to prevent an allergic reaction. Unfortunately, pharmacies around me opened late that morning, so I had to rush from pharmacy to pharmacy to secure the tablets. Once the emergency was handled, I rushed in the office, and my manager greeted me with a "we have to talk" look. He explained to me that there are rules in this company and one rule is that I had to be in by 9:30 a.m. Obviously I said sorry for being 20 minutes late, because I acknowledged violating the rules. I also added that there was a family emergency and that I had informed him as soon as I found out I would not make it. I received little to no understanding, because "rules are rules." Back then, I was certain this manager would not last long in the company, because even though rules are rules, not addressing a life-threatening emergency situation is a crime, and companies have no authority over their employees to request them to commit a crime.

Contrast this to another company, where employees arrived between 8:00 a.m. and 2:00 p.m. and did their 8 hours. The last person closed the office around 11:00 p.m. I took 3-hour-long lunch breaks several times and distributed my work in two chunks. As long as the work was done, everyone was happy. Flexibility trumps rigor as long as it provides everyone a win-win situation.

As a freelancer, I have experienced that I am often hired to finish projects, not to get paid by the hour. Project-based pay may help both parties if it is done right. Freelancers may choose to work in the way they want to, while clients can plan their budgets. If you can offer something unique and become the top expert at a niche, you can reap disproportionately big rewards.

Finding your own niche is often more lucrative than climbing a corporate ladder. This is why your career plan and your learning plan are tied together. It is not a job that can provide you with security. It is rather your income earning skills that help you create a ladder for yourself. This is why one of the best investments you can make is learning a skill that can be sold on the market.

Job security is also becoming less significant than in the past. While elite developers tend to find jobs easily, in the age of venture capital, externally funded startups hire and fire their developers easily. Established, publicly traded companies may lay off employees to help their shareholders realize profits and drive the share price up. It is not a wise move to put all your eggs in one basket and rely on just one company to provide a long-term career plan for you even if most good companies help you plan your career and develop your skills. It is simply less risky to provide job security for yourself by learning skills that help you make a living even if you get fired. A skill can be that you are a genius problem solver with a track record of having worked for Amazon and Google in responsible positions.

Obviously, you will not have a problem with finding a job even at the toughest times, because you can deliver something unique. Job security at these companies also tends to be higher, because even if there are layoffs, chances are your job won't be affected.

However, most software developers are in a different position when it comes to risks and exposure to getting fired. Therefore, learning how to find multiple clients for yourself and how to make money freelancing or how to create your own products and services may pay off more than investing time and effort in climbing your corporate ladder. Sure enough, there is nothing wrong with becoming a great expert, manager, or intrapreneur. Some of your skills are even transferable. However, other aspects of your career capital only apply to the company you are working for. If you change your job, some fraction of your career capital will be taken away from you.

This is where your personal brand and your long-term connections come into play. The easier you can get new clients from the market, the more secure your position is on the job market. In the 21st century, this is achieved through providing products or services others buy.

A few years ago, a colleague of mine referred to his services as gigs. He made five figures on the side doing his gigs as a designer and had many clients. He then asked us to reduce his working hours, and within a year, he resigned and went self-employed. He never looked back ever since.

From the demand side, companies also have trouble with the concept of full-time employment. Similarly to software development, Single Point of Failure (SPOF) threats exist in positions that cannot be substituted by anyone else. If a person in a SPOF position resigns or becomes unable to work, the company may risk business continuity.

Another problem with full-time employees is that they have to be paid even if they have nothing to do. I have been instructed as a manager several times throughout my career to "keep some people busy" with tasks that develop their skills and make them stay with the company, simply because their skill was not needed in business-critical projects. However, getting a full-time employee from the job market took us months and cost us more than $10,000, so it paid off for us to keep the full-time employee and use his time on something creative.

More than 8 years ago, I already experienced running a remote team hiring freelancers from the market to perform short-term gigs. The hourly rate of these freelancers wasn't a lot higher than that of full-time employees, and they got the job done. Sometimes offshore or nearshore labor gives you the same or better skills than what's available to you locally, and if you work in a society with high wages and high costs of living, outsourcing your job pays off big time. The only problem was the management overhead; however, as I gained more experience documenting repetitive tasks, I managed to automate most of the process. Initially skill level of freelancers varied, but after building a network of reliable professionals, the initial investment paid off.

I can also recall a story one of my colleagues told me about inequalities. He worked for a company with multiple branches. Once an employee was transferred from Switzerland to Hungary. Twenty years ago, it was not uncommon to earn $12,000 per year. For the same job, the Swiss employee was rewarded more than $120,000. As the transfer was made, the Swiss employee retained his whopping $120,000 salary, and he was proven to be less effective at work than my colleague.

Employees earning lower wages are often fed the explanation that their contribution is worth less, because their work is less valuable in a less resourceful system.

Indeed, there is some truth to this explanation: taxes are different in each country. Some countries may provide more than 30 days of holiday. Some countries may provide mandatory social security contributions costing both the employee and the employer more than 400 dollars of extra costs a month. Some countries force employers to finance a month of sick leave every year, as well as getting exposure to a risk of maternity and paternity leave ranging up to a year per child. Some countries also have high taxes. Some countries force you to implement expensive data security frameworks like GDPR, and some countries force you to form a labor union as soon as your number of employees reaches 100. Some countries make it illegal to fire employees even if they don't do anything productive, as long as they do not violate the law. Some countries have a complex legal system, where it costs you a high four-figure sum just to examine if you are allowed to give remote work opportunities for an employee.

Chances are countries with the preceding features make it harder for employers to provide full-time employment opportunities, and these constraints would decrease wages. However, in reality, many welfare states operate with many of these rules, and they have some of the highest salaries available to IT professionals. Therefore, there must be a force that is more significant than these constraints. Indeed, this force is nothing else but demand and supply on the market. If people are willing to work for $1,000 per month in a country, companies will pay $1,000 per month even if clothing costs them the same as in countries where their wage is $10,000 per month for the same job.

As real estate rental prices go up, wages also increase. As taxes or general costs of living increase, wages also increase. As a financial crisis eliminates job opportunities, wages decrease, because demand stays the same, while supply decreases.

■ **Note** With information spreading on the Internet, skilled workers have realized that wage gap between countries is not primarily because of utility of work, but because of market forces. Therefore, there is an evident brain drain from countries providing less resources for work toward countries that pay higher wages. As remote work and freelancing opportunities are spreading all over the world, this brain drain does not even require skilled labor to travel and bear the burden of higher costs of living and higher taxes. Alternative lifestyle options such as working from home or not even having a home and becoming a digital nomad are spreading.

The freedom of choice in the 21st century enables some people to think outside the box and choose an environment where they can thrive. Barriers such as a 9–5 mentality for keeping knowledge workers in a small comfort zone, full-time employment, and corporate ladders are being abandoned by the masses in favor of other opportunities. In fact, there is a trend from many governments to simplify and decrease the tax and administration burden on freelancers and small businesses. The incentive from the government's end is not to make you contribute less back to the social welfare system, but to make you grow further so that you can create jobs.

Note Internet marketers tend to spread intellectual fog around employment by demonizing office work. The same Internet marketers often go to cafés or co-working spaces to get their job done, because they cannot force themselves to be productive at home or on the beach. If you are working in an office, chances are you are enjoying a lot of benefits in the office that not only make you productive and focused but also help you be more satisfied and separate work from leisure. Offices are neither good nor bad; they exist for a reason.

I can still recall the time when I was doing remote work from home. I never had to go in any offices. Once in a gloomy windy evening, I was jogging, and I saw stylish offices with well-dressed employees staying busy in the late afternoon; and that was when I realized that I actually miss this environment. At that time, I didn't have the skills and knowledge to earn more money as an employee in an office. While working from home means freedom for many people, it was nothing else but lack of resources and lack of social connections for me, because I had to spend my work time alone and I also had to work more than 8 hours a day under these conditions to make the same amount of money.

Although governments around the world tend to favor employment, more and more employees are clearly fed up with getting second-rate treatment because of their location. Since 2016, my mission has been to provide alternatives to these software developers in the form of locating optimal remote work opportunities, encouraging remote applications, turning employees into freelancers, and turning employees into creators of products and services.

I regard myself as a pioneer in terms of making my own path and sharing the results of my work with others through my books, courses, and private mentoring programs.[1] Some surprising things I have done in my own career are also available to others who put in the effort. This includes the following:

---

[1]http://devcareermastery.com/coaching-program/

- Finding freelance opportunities for \$100–\$250 per hour[2]

- Finding projects for \$5,000 that you can complete in 50 to 150 hours depending on your experience[3]

- Securing jobs in two countries, going through the interview process remotely

- Enjoying 300 days of sunshine a year, living and working next to the seashore

- Securing a remote work opportunity from the comfort of my home and working remotely for a year

- Tripling a starting salary at the same company while securing bonuses

- Becoming a development manager at multiple companies

- Earning a consistent stream of income from royalties arising from both self-publishing books and courses and cooperating with publishers

- Creating two blogs that allowed me to earn five figures a year on the side in a bit more than a year[4]

All information is available to you to accomplish these points and more. Whether you get this information from me or from other sources, it does not matter. What truly matters is that you take action. Good luck!

---

[2] On top of your hourly rate, you have to calculate the cost of marketing yourself and keeping in touch with your clients. You also have to factor in some preparation for a highly paid gig such as teaching a webinar for \$250. Even though you are paid to host a webinar, you have to create the slides for it, which might take you another 1 to 2 hours, reducing your hourly rate. However, once you create your slides, you can reuse your content for other webinars, eventually making the \$250 hourly rate manifest in reality.

[3] I did two of these projects in 2018 while I was working full-time. I agreed with my client on a relaxed timeline and worked on these projects 10 hours a week.

[4] Actually, I am not that proud of this accomplishment, because I could have doubled down on personal brand building in 2016–2017 at the expense of gaining career capital as a professional. However, I chose not to go down that path, because I have seen where personal brand building leads to in the case of people who leveraged the opportunity of building an online presence to escape from work. I believe in "moving toward" motivation, and escaping from work does not motivate me. Maximizing career capital does.

Redefining Education in the 21st Century

Disclaimer My opinion is limited to my experience, and I graduated in the area of STEM: Science, Technology, Engineering, and Math. There are some university programs that violate the law of demand and supply and produce degrees that make close to zero sense from an economic perspective. There are some university programs that present pseudosciences in a scientific way and certify people with status that is not meant to be questioned even at court. If a suspect is convicted, they go to jail. If it turns out the expert makes a mistake, they are forgiven. This section neither formulates opinion on these phenomena nor claims that anything presented applies to these subjects. Our scope here is solely STEM.

First of all, I would like to express my gratitude toward the republic of Hungary[5] and the Budapest University of Technology and Economics for providing me with free education to obtain an MSc and finance my involvement in two research projects and helping me get started in my career.

As a former student, I feel for people who accumulate six digits of debt to finance their studies. I never had this debt, and I feel privileged for the opportunity to get my career started.

Opposed to many YouTubers, authors, and Internet marketers, I am not burying the traditional education system; and I don't tell you urban legends about why you are better off quitting college to make your path.

Note Recall the concept of intellectual fog. While traditional education is by far not perfect, the incentive of Internet marketers showing you their Lamborghinis, luxury penthouses, expensive suits, or vacations is that you buy their products. While some Internet marketers have a very high IQ and many of them believe what they teach, they rarely have the life experience or the mindset to appreciate the positive aspects of traditional education.

I didn't go to a university because my parents expected me to. It was my own expectation toward myself, and I consciously prepared for this path since the age of 12. I had an IQ around 120–130, and I was a top 5% student, but most of my results came as a result of hard work. I was never gifted in any way with talent that allowed me to skip classes or attend a university at the age of 14. I chose the traditional route and focused on enjoying the process and absorbing knowledge.

---

[5] Back then, republic was included in the name of the country.

Note The decision of whether it makes sense for you to go to college or university depends on several factors: your goals, your lifestyle, your current life situation, and your current skills and connections. I consciously exclude the expectation of other people, including the expectations of your parents from this list. For some people, it does not make sense to attend college at all. For others, it's the best start of their career. For others, it's optimal if they start their career and then they attend college or university once they make enough money.

The following benefits apply to colleges and universities:

- **Immersion:** These institutions specialize in training students. You are exposed to a deep level of knowledge that enables you to outperform most self-trained bootcamp students.

- An opportunity to find your niche by getting access to lab equipment and other resources and trying yourself out in several areas of IT engineering and management.

- **Connections**: You build your career capital by getting to know people. Some of them will become CEOs, others become self-employed, and others become skilled employees.

- A low-risk environment to learn how to perform under pressure, fail, earn your stripes, and become more resilient.

- Some hidden gems in forms of classes, where the instructors break out from the barriers of traditional education and innovate.

- **Social acceptance and status**: Let's face it, whether you think it is good or bad to fit in, it definitely does not hurt your chances.[6] Especially in formal environment, the market may reward formal education with higher wages.

The opportunity cost for these benefits is that it takes time and money. The time to finish the curriculum rarely makes it a financially optimal move in the short run.

---

[6] Exceptions apply. For instance, in business, a PhD may slightly hinder your connections with self-made business partners having no formal education.

Note If you know exactly what you want and what skills you need and you take full responsibility for establishing yourself as a self-employed person or an entrepreneur and your need for certainty is low, then it is an option to learn the skills you need with alternative education. Start your business, and make money faster than your peers who choose traditional education.

I admit, when I was 18, I had a high level of need for certainty and security. Finishing high school and starting work was a scary thing for me. Although there are more opportunities today than in 2001, even now, with my 18-year-old mind, my thought process would just not have been mature enough to establish myself and make my path.

Note If I had a degree in IT engineering and I wanted to retrain myself in marketing, sales, or business administration, unless I chose a prestigious MBA program to build connections, chances are I would just use alternative education to retrain myself. Similarly, I have interviewed self-taught lawyers, teachers, and philosophers, who had a degree in their profession, but didn't find it worthwhile to attend an IT engineering college.

Unfortunately, the ride at my university was rough, because sometimes I had to learn some subjects I never appreciated and knew I would never use. This is one dark side of higher education. Therefore, if your goal is to earn a skill that enables you to make money as fast as possible, college and university is not the fastest way to go. The question is what fraction of the population is ready for taking this level of responsibility at the age of 18.

Note The biggest threat to young, inexperienced students is to submit to learning intellectual fog and try out get rich quick schemes that never meant to make you rich. While failures give you experience, the speed of gaining career capital is often lower than using accredited sources of education.

Another drawback of higher education is that it mainly prepares you to become an employee. The emphasis is on the word mainly. As I looked for alternative opportunities, I can recall at least five of my university professors who were either self-employed or ran a business themselves. I had access to their classes, and they enabled me to learn from their past failures and successes, which gave me a solid foundation on preparing me for what was to come. These opportunities are available to many students, but only those benefit from them who open their eyes and ears to locate these opportunities and specializations.

The other extreme is about important university subjects that stay outdated. I can recall one specific professor who was very successful in the 1980s and never updated his knowledge. PhD students in his team learned what he knew, and no one innovated that subject. It was sad to see that free information on the Internet was at least 10 or often 20 years more recent. In the world of IT, by the time you graduate, trends in knowledge change.

For instance, I never learned about microservices in university. I did learn about Service Oriented Architectures only by sneaking in a lecture I didn't have to take.

I never learned about client-side rendering, and I thought JavaScript was a toy language enhancing what's possible with just HTML and CSS. Since then, I have made a career out of learning and mastering JavaScript.

Therefore, universities are rarely good for learning cutting-edge stuff.[7] They are great for giving you the fundamentals that endure the test of time, allow you to build connections, and help you get started. It will be your responsibility to keep yourself updated.

Traditional education is challenged in many fronts in the 21st century:

- Providing specialized courses for the fraction of the cost of universities where you learn everything you need and nothing you don't.

- The emergence of software developer bootcamps, sometimes with job guarantees.

- Widely available mentorship programs that are still cheaper than university education.

- Education specialized in helping you learn marketable skills not in connection with employment.

- Free or cheap learning resources available online.

- Paid courses have drastically leveled up in quality; and the best resources are now using advanced technology, learning management systems, and even AI to provide you with the best possible source of education.

- MOOCs (Massively Open Online Courses).

---

[7] Exceptions apply. One of my peers from university committed to an IBM lab and earned a starting salary of around $50,000 after graduation in Hungary, where others were happy to earn $15,000 a year (2006 data).

- Certification institutions often breaking down the barriers of entry.[8]

- Personal branding opportunities that substitute the need for a degree.

- The ability to take university courses remotely and collect credits for either a fraction of the price of higher education or for free. You can get a college or university degree in 2019 by absorbing mostly online resources and paying a fraction of the tuition fee. You can tailor the subjects you need getting rid of mandatory subjects universities give you and set the pace for yourself, allowing you to work while you study or allowing you to obtain your degree faster than regular students.

Most of these changes happened in just one generation; therefore, your parents might have never told you about these opportunities. They are still there. This is why my mission is to contribute to the democratization of education and providing a continuously increasing quality level that enriches education alternatives.

Does this mean the end of the traditional education system?

Unlikely.

Systems tend to be challenged and disrupted. However, following Nassim Taleb's concept of antifragility, these disruptions tend to make systems stronger, because as they heal from damages, they become stronger than before.

Just look at www.coursera.org/. It provides a disruptive way of education by giving you a free start and cheap options to complete online courses provided by the world's best universities, and you get an ability to earn degrees.

If I had a parent or some friends who told me that I would be more than fine if I took these courses seriously or I had an older brother or some online role models who completed these courses online, chances are even at the age of 18, I would have found taking responsibility for my own education a lucrative alternative to attending university. The change our world is going through is that more and more credible role models emerge and you get free access to them online.[9]

---

[8] For instance, research the Scrum certification offered by the Scrum Alliance and Scrum. org. One provides you with an easy test for obligatory expenses of more than $1,000 and a regular renewal fee, while the other allows you to complete a hard test in exchange for $150 (2019 data).

[9] Emphasis is on the word credible, because obviously, scams also surface. Some scammers present themselves as an alternative to traditional education and justify a high price tag. Some of the most damaging and deceptive programs have been shut down by the American government.

Looking at Coursera, traditional education is not meant to compete with these sources. It is meant to cooperate with them. After all, which institution has more resources and government funding to provide education than colleges and universities? Where do the most prestigious courses come from? Cooperating with colleges and universities is a win-win situation.

Unfortunately, these changes are slow, and we will experience more disruption that makes it likely for the general public to question the utility of traditional education.

Final Notes

In the last two sections, we have concluded that it is not necessary to attend a college or university to obtain a degree or relevant knowledge and it is not necessary to go to an office to be a productive and appreciated part of a software development team.

Hopefully, you are now convinced that there are a lot of opportunities out there to improve your current situation. You have received the tools and strategies to plan your next 5 years. When executing your plans, you will need to work on your thoughts and emotions to execute your steps smoothly. Fortunately, emotional intelligence gives you a great toolbox to become your best self and face any obstacle in your way.

You will need personal branding and an online presence, and you will have to learn how to write application packages. Chapter 4 gave you all information you need to make this happen.

Once you get hired, you will have to make a lasting impression with your professional attitude and communication skills. In Chapter 5, we dealt with tools and techniques that help you present your achievements and communicate your interest, taking the interest of others into consideration. You don't need these skills for establishing your online presence, but you do need all these skills to negotiate a raise or get hired. This is why Chapter 5 was between your online presence and the chapters on negotiating raises and interviewing. These two chapters build heavily on the first five chapters, as your professional attitude, your accomplishments, your emotional balance, and your communication skills will back your quest up.

You now got all the building blocks to double your career velocity. Use them wisely, and remember it is always better to do things right than just maximizing short-term monetary rewards. Indeed, you can achieve fast results, but remember responsibility does not follow money. It's the other way around. Money almost always follows responsibility. Choose maximum responsibility, and new doors will open for you either with your current employer or elsewhere.

If you have a success story to share, or you get stuck, feel free to get in touch with me at zsolt@devcareermastery.com.

I wish you all the best in your career, take care!

I

Index

© Zsolt Nagy 2019
Z. Nagy, *Soft Skills to Advance Your Developer Career*,
https://doi.org/10.1007/978-1-4842-5092-1